A C S   S Y M P O S I U M   S E R I E S **369**

Good Laboratory Practices
An Agrochemical Perspective

Willa Y. Garner, EDITOR
U.S. Environmental Protection Agency

Maureen S. Barge, EDITOR
FMC Corporation

Developed from a symposium sponsored
by the Division
of Agrochemicals
at the 194th Meeting
of the American Chemical Society,
New Orleans, Louisiana,
August 30–September 4, 1987

American Chemical Society, Washington, DC 1988

Library of Congress Cataloging-in-Publication Data

Good laboratory practices: an agrochemical
perspective: developed from a symposium
sponsored by the Division of Agrochemicals
at the 194th meeting of the American Chemical
Society, New Orleans, Louisiana, August 30–
September 4, 1987 / Willa Y. Garner, editor;
Maureen Barge, editor.

    p.   cm.—(ACS symposium series, ISSN 0097–6156; 369)

Includes bibliographies and indexes.

ISBN 0–8412–1480–8

    1. Chemical laboratories—Quality control—Congresses.
2. Chemical laboratories—Standards—Congresses.

    I. Garner, Willa Y., 1936–   .
II. Barge, Maureen, 1947–   . III. American
Chemical Society. Division of Agrochemicals.
IV. American Chemical Society. Meeting (194th: 1987:
New Orleans, La.)   V. Series.

QD51.G66   1988
542'.1'0289—dc 19   88–6330
                                          CIP

ACS Symposium Series

M. Joan Comstock, *Series Editor*

1988 ACS Books Advisory Board

Foreword

The ACS SYMPOSIUM SERIES was founded in 1974 to provide a medium for publishing symposia quickly in book form. The format of the Series parallels that of the continuing ADVANCES IN CHEMISTRY SERIES except that, in order to save time, the papers are not typeset but are reproduced as they are submitted by the authors in camera-ready form. Papers are reviewed under the supervision of the Editors with the assistance of the Series Advisory Board and are selected to maintain the integrity of the symposia; however, verbatim reproductions of previously published papers are not accepted. Both reviews and reports of research are acceptable, because symposia may embrace both types of presentation.

Contents

Preface

PROPOSED GENERIC GOOD LABORATORY PRACTICES have been published in the *Federal Register*; there will be a 90-day public comment period. This volume is intended for the chemists, quality assurance personnel, and laboratory managers who will need guidance in implementing the good laboratory practices for their studies. This book takes a major step toward a united effort to ensure that all studies intended for support of a pesticide registration are in compliance with good laboratory practice standards.

The symposium on which this book is based fostered an understanding of the various aspects of developing or improving a quality assurance program for chemistry studies. Designed to bring together chemists and quality assurance specialists from industry, academia, and state and federal governments, the scope of the symposium ranged from summarizing current practices and identifying probable changes to defining what needs to be done and how to do it. The program focused on the cradle-to-grave philosophy of monitoring a study. The presentations began with an overview of good laboratory practice regulations from the perspectives of government, primarily the U.S. Environmental Protection Agency, and industry. The overview was followed by a discussion of the role of management and the interactions required between bench chemists and the quality assurance unit. The program then continued with the "hows" and "whys" of implementing the regulations to chemistry studies.

As organizers of the symposium and editors of this volume, we thank the contributors, whose expertise and generosity with their time will make this book a valuable reference for those working in the quality assurance field. We also wish to express our appreciation to the National Agricultural Chemicals Association for their interest and support and to the Division of Agrochemicals of the American Chemical Society for sponsoring the forum.

WILLA Y. GARNER
U.S. Environmental Protection Agency
Washington, DC 20460

MAUREEN S. BARGE
FMC Corporation
Princeton, NJ 08543

January 20, 1988

Chapter 1

Good Laboratory Practices

Birth of a New Profession

Carl R. Morris

International Center for Health and Environmental Education,
4600 Pinecrest Office Park Drive, Alexandria, VA 22312

A brief overview of some of the historical milestones
in national and international Good Laboratory
Practice (GLP) are presented. In particular, the
work of the U.S. Food and Drug Administration and the
U.S. Environmental Protection Agency to develop
national GLP regulations are discussed as well as
their efforts, within the Expert Group on GLP of the
Organization for Economic Cooperation and Development
(OECD), to harmonize GLP guidance for the 24
countries of this international organization. The
Expert Group was able to develop an international GLP
guidance document on the Principles of GLP as well as
two other guidance documents relating to the
"Implementation of OECD Principles of GLP" and "OECD
Guidelines for National GLP Inspections and Study
Audits." The advent of national GLP regulations and
international guidance on GLP has resulted in the
creation of a new scientific, managerial
professional--the quality assurance unit manager.
The responsibilities of this new professional are
discussed as well as the challenges that this
professional will face in the future.

The issue of the quality of laboratory data being submitted to
governmental agencies is a major concern of the public as well as
state and federal regulatory agencies. These concerns have resulted
in the implementation of administrative procedures by regulatory
agencies to assure that submitted data is reliable and of the
highest quality with the present state-of-the-art. In order to
better understand how these good laboratory practice (GLP) concerns
were addressed, a brief overview of some of the historical aspects
of GLP implementation is discussed including the birth of a new
profession--the quality assurance unit manager.

0097–6156/88/0369–0001$06.00/0
© 1988 American Chemical Society

Historical Perspective

During the period from late 1960's-1975, federal regulatory agencies were faced with a number of discrepancies in data submitted to them. There were instances of laboratories not following protocols, the lack of documented standard operating procedures (SOPs) and, if available, poor compliance with SOPs. Several laboratories had a general problem of poor documentation and incomplete reporting of data to regulatory agencies. It was clear that a better job needed to be done in the management of developing and reporting laboratory studies. In some cases, data were submitted to regulatory agencies which subsequently established that the data submitted were never developed in any laboratory. To respond to these issues, Congress urged regulatory agencies to enact regulations to address these problems. The U.S. Food and Drug Administration responded with a proposed Good Laboratory Practice (GLP) regulation in 1976 followed by a final regulation in December 1978. The Environmental Protection Agency joined in these activities with their initial GLP proposal in May 1979 and a final regulation in November 1983.

Recognizing the importance of these GLP regulations on the international chemical trade and their potential as non-tariff barriers to trade, the EPA and FDA joined with other countries to address these issues on an international basis. Since 1977, the U.S. as well as the other 23 Members of the Organization for Economic Cooperation and Development (OECD) have been involved in extensive international consultations concerning harmonization of chemical programs. As a part of these efforts, an international Expert Group on GLP was established in 1978. During the next 3-4 years, this OECD Expert Group on GLP undertook a major effort directed toward the development of international guidelines for Good Laboratory Practice (GLP). The principle objective of these guidelines was to assure, to the extent practicable under the laws of the OECD member countries, that data developed to meet one country's requirements would be acceptable to other countries. There was strong endorsement of the work of the OECD Expert Group on GLP at meetings of high level national regulatory officials in May of 1980, and in November 1982. In May 1981, OECD member countries adopted a formal decision on the mutual acceptance of data which, to the extent practicable under the laws of OECD member countries, binds member countries to accept data generated according to the OECD Test Guidelines and the OECD Principles of Good Laboratory Practice for assessment purposes.

In addition to the development of the OECD Principles of GLP, the OECD Expert Group was given the responsibility of developing two additional guidance documents--one for the Implementation of OECD Principles of GLP and one as OECD Guidelines for National GLP Inspections and Study Audits.

The Implementation document encourages member countries to adopt the OECD Principles of GLP into their legislative and administrative frameworks. As a part of the adoption and implementation process, national authorities should document their compliance programs, including provisions for the declaration on the part of each laboratory that the study conducted therein was in accordance with the OECD Principles of GLP or with national regulations or equivalents conforming to these Principles. National compliance programs should utilize laboratory inspections and study audits as principal mechanisms whereby they can monitor compliance to the Principles of GLP. It was further recommended that national authorities utilize properly trained personnel who are competent to assess the compliance of laboratories with the Principles as well as to administer the GLP compliance programs. Within the documentation of each national GLP compliance program, there should be provisions for actions which may be taken by the national authority for non-compliance with the Principles and provisions to remedy any deficiencies.

With respect to international recognition and cooperation, the Implementation document identifies the need for an international mechanism for recognizing the comparability of GLP compliance programs of each country. Although bilateral consultations and bilateral memoranda of understanding between competent authorities have provided useful guidance in the past, it is recognized that a multilateral mechanism for recognizing and fostering the development of comparable national GLP compliance programs is a more resource efficient approach. I, personally, support this type of an approach. Although we may have some individual problems that are unique to our respective national laws, I believe that the major GLP implementation issues are essentially the same from one country to another. I believe that we should strive to identify those common issues and share the information concerning the approach to and resolution of GLP implementation problems. The OECD GLP Implementation document encourages international consultation and verification of GLP compliance programs. It supports the establishment of an international GLP forum in which national competent authorities could meet at least once a year to (1) discuss technical and administrative matters arising from their respective national GLP compliance programs, (2) promote cooperation between competent national authorities, (3) exchange information on the training of inspectors, and (4) promote for inspectors, periodic seminars dealing with the provisions of the Principles and the various aspects of inspections and study audits.

The OECD Guidelines for National GLP Inspections and Study Audits document serves as a companion document to the Principles and is intended to provide national authorities additional guidance in preparing and implementing their national GLP compliance programs.

As a result of OECD Council action, the major points recommended by the OECD Expert Group have been endorsed. The Council noted that member countries will establish their compliance procedures progressively according to their respective national priorities. The Council instructed the Environment Committee and the Management Committee of the Special Program on the Control of Chemicals to:

(1) foster direct communications between national authorities and to provide a forum within the organization to discuss technical and administrative matters related to GLP compliance procedures; and

(2) pursue a program of work designed to facilitate the implementation of these recommendations with a view toward member countries developing bilateral and multilateral arrangements for the mutual recognition of national GLP compliance procedures.

National GLP Implementation Programs

Over the past several years, several countries have been involved in developing national GLP regulations/guidelines and in the implementation of GLP compliance monitoring activities. A brief overview of those activities occurring here in the U.S. is presented.

U.S.

The most progress toward the implementation of GLP regulations for non-clinical studies and GLP compliance monitoring programs is best exemplified by the U.S. Food and Drug Administration's (FDA) efforts over the past 10 years. The FDA GLP regulations were published as final rules in 1978. Its GLP Compliance Monitoring program has been actively involved in inspections and study audits both domestically and internationally for submitted studies. In addition, the U.S. Environmental Protection Agency has proposed GLP regulations for health and environmental studies under the Toxic Substances Control Act (TSCA) and for health studies under the Federal Insecticide, Fungicide and Rodenticide Act (FIFRA). Final GLP regulations under both of these authorities was published in November 1983. Both TSCA and FIFRA regulatory programs have implemented inspection and study audit activities involving national and international testing laboratories. The health inspections are coordinated through an interagency agreement with the FDA Compliance Monitoring program whereas the environmental inspections are being conducted by EPA Headquarters and Regional Inspectors. Study audits for all studies are being conducted by EPA Headquarters technical staff. These activities represent a major coordination effort on the part of U.S. regulatory agencies to harmonize this country's regulatory initiatives in GLP.

The Birth of a New Profession

As previously indicated, these GLP regulations and international guidelines were written to address the issue of the conduct of studies and assuring their quality. In order to address these issues, each of the regulations and the international guidelines calls for the establishment of a Quality Assurance Unit or "quality assurance function" within each laboratory. They also specify that certain tasks be carried out by this unit or function. These requirements have resulted in the creation of a new scientific, managerial professional – the quality assurance unit

manager. Unfortunately, many individuals were given responsibility
for these new tasks with no guidance, other than the regulations, on
how to conduct such activities. The regulations addressed broad
areas of responsibility but the detailed implementation was left to
these "new process managers." As federal inspectors began to arrive
at laboratories to evaluate how the laboratories were doing in
complying with these new GLP regulations, the quality assurance unit
manager became heavily involved with regulatory affairs issues,
management of processes to which they had no direct control, and, in
some cases, outright hostility from study directors. These new
professionals were totally unprepared for such challenges. Over the
course of the last 10 years, these professionals have been able to
win the confidence of their technical peers as well as those of
federal regulators through a long and arduous path filled with trial
and error, persistence, long hours, and a rare, thank you. However,
new problems are now facing this profession.

During the past two years, a considerable number of quality
assurance unit managers have left the profession or have left their
management position in quality assurance units of several major
laboratories. Their departures have left these facilities with
relatively new, inexperienced young professional managers. Although
the FDA Good Laboratory Practice (GLP) regulations have been in
place for nearly 10 years and the EPA GLP regulations for 4 years,
training programs for these new professionals for addressing the
issues of GLP compliance have been offered by only a limited number
of training institutions. Academic training programs in the sciences
have only touched on the regulatory affairs issues now being faced
by these new practicing professionals.

The impact of these issues on the regulatory compliance process
may be significant. Regulatory agencies are looking for laboratory
organizations that are stable with a high level of quality and
integrity in its staff. Fewer regulatory visits will be required in
those laboratories that demonstrate a highly trained, competent
staff with a history of institutional continuity. As the scope of
studies requiring good laboratory practice management requirements
is expanded, the responsibilities of these new quality assurance
professionals will increase as well as the demand for sufficient
numbers of competent, well-trained professionals. One of my
personal goals over the next 5 years is to help these new
professionals prepare themselves for their new challenges as
laboratory quality assurance managers.

References

1. Final Rule for Good Laboratory Practice Regulations under the
    Federal Food, Drug, and Cosmetic Act. 21 CFR Part 58. Federal
    Register, 43: pp 59986-60025, December 22, 1978.
2. Good Laboratory Practice in the Testing of Chemicals, Final
    Report of the Group of Experts on Good Laboratory Practice, No.
    42353, Organization for Economic Cooperation and Development: 2
    rue Andre-Pascal, 75775 Paris Cedex 16, France, 1982.

3. Final Rule for Good Laboratory Practice Standards under the
    Federal Insecticide, Fungicide, and Rodenticide Act (FIFRA). 40
    CFR Part 160. Federal Register, 48: pp 53948-53969, November
    29, 1983.
4. Final Rule for Good Laboratory Practice Standards under the
    Toxic Substances Control Act. 40 CFR Part 792. Federal
    Register, 48: pp 53922-53944, November 29, 1983.
5. Morris, C.R. OECD Update. J. Amer. Coll. Toxicol., 5: pp 293-
    296, 1986.
6. Morris, C.R. Quality Assurance Principles for Applied
    Toxicology. In: Safety Evaluation: Toxicology, Methods,
    Concepts and Risk Assessment; Melman, M.A., Ed.; Advances in
    Modern Environmental Toxicology Series No. X; Princeton
    Scientific Publishing, pp 117-126, 1987.

RECEIVED March 21, 1988

Chapter 2

Industry Perspective on Good Laboratory Practice Regulation of Chemical Studies

John F. McCarthy

National Agricultural Chemical Association, Washington, DC 20005

In spring 1985, NACA began addressing GLPs for chemistry studies, and formed a Subcommittee to prepare
guidelines for use by member companies. The Subcommittee deliberated, consulted EPA and completed the
final document in mid-1986. The guidelines, modeled
after FDA, EPA and OECD GLP regulations for animal
studies, address residue (laboratory and field), metabolism (plant and animal), and environmental chemistry
studies done for FIFRA registration requirements.
However, they are more general due to the breadth of
chemistry studies involved. During the development of
GLP regulations by EPA, NACA encourages "non-compliance" audits of companies to assist them in their GLP
programs. This provides EPA and industry opportunity
to understand differing stances prior to a compliance
situation. These experiences should aid EPA in
developing regulations and prepare industry for
regulatory compliance. Compliance should be phased in
so that completed and ongoing studies are accepted
even if regulations are not precisely met.

Industry's commitment to quality science is fundamental. The
practice of quality science is at the heart of our industry. It is
essential that the scientific community, regulators and the public
have confidence in what we do and how we do it. In other words,
there should be no doubts about the validity of our data and the
competence and integrity of our scientists.

We all want to be trusted - it's a fundamental human need.
Trust is something one earns. A basic principle of science is that
experiments should be repeatable. That is, one investigator should
be able to repeat the work of another. If experiments can't be
repeated, then the trustworthiness of the original investigator may
come into question. However, we cannot count on duplication as the
only method of verification. There are simply not enough resources
to do that. In addition, it would not be a wise use of resources to
verify everything done by the industry by repeating the experiments.

0097–6156/88/0369–0007$06.00/0
© 1988 American Chemical Society

There has to be another way. That's where GLPs and laboratory
audits come into play.

The incident which occurred at a major toxicology testing
laboratory in the middle 1970's shook up the industry and taught us
a lesson. Not only did we learn that we must pay more attention to
what others (contractors) do for us, but we also needed to develop
more rigorous GLP procedures within the corporate laboratory. This
incident led to the promulgation of regulations by the Food and Drug
Administration (FDA) on Good Laboratory Practices for non-clinical
laboratory studies in 1978. This was followed by EPA with GLP
regulations for the same kinds of studies in November 1983. I think
it's safe to say that back in 1976 when the FDA regulations were
first proposed, they weren't welcomed with open arms by the
industry. The impact of these were primarily on the pharmaceutical
industry. There was a lot of verbal sparring. The proposal was
attacked as unnecessary, prohibitively expensive, and conducive to a
stifling bureaucratic blanket on creative research and scientific
investigations.

By the time the EPA regulations were proposed in 1980, the
environment had changed somewhat. Industry had learned the value of
GLPs. This is not to say there were not thoughtful and germane
comments on the EPA proposal. I believe all of this experience has
prepared us for the GLP regulations of chemistry studies. However,
before I get into the details of this aspect, I would like to give
an overview of some of NACA's activities with respect to GLPs.

NACA Activities

Starting in about 1978, the Research Directors Committee of the
Association began to address the subject. The membership of the
committee was polled regarding their views on data retention. This
was largely stimulated by proposed EPA guidelines on this matter. I
won't attempt to summarize the results of that survey, but emerging
from it was a consensus definition on the term "raw data," and a
consensus statement on retention of samples. The definition of "raw
data" which the industry preferred was as follows:

"The term 'raw data' means laboratory or field worksheets,
records, notes, or memoranda that are the result of observations,
measurements or other activities which contribute significantly to
the conclusions drawn from the testing or evaluation of a pesticide
for purposes of registration."

The words "contribute significantly" and "for purposes of
registration" were key, and represented an important change in the
definition vis-a-vis what was proposed by EPA. It was felt that the
inclusion of these words would be a more reasonable definition
because it reduced the scope of raw data retention to the very type
which was needed for validation and which could be reasonably
expected to be retained.

With respect to retention, it was the consensus of the
Committee that EPA's definition was reasonable. However, it was
apparent that there was a variety of sample retention practices
within the industry. Most companies did not make a special effort

to retain samples of all specimens used in all of the testing at the time of the survey. This has now changed.

Emerging from this exercise was the development of official NACA position papers on "The Reliability of Test Data in Pesticide Research" and "Good Laboratory Practices" in 1979. Needless to say, the policy on the former indicated that the industry supported the concept of reliable test data in pesticide research. The issue was how does one determine reliability or validity. The industry addressed the question this way:

"Tests of validity should be based upon the appropriateness and quality of the test design; the manner in which the research is executed; documentation and records to support scientific logic and expertise exhibited in the evaluations and conclusions reported."

As you can see, this touched upon some of the principles of GLPs, namely, documentation, records and expertise. With respect to the Good Laboratory Practices position paper, NACA indicated the need for consistency of regulations among the various regulatory agencies. The paper went on to recommend that the FDA GLPs be uniformly adopted by all regulatory agencies for regulation of health effects testing. The Association also endorsed the concept that all GLP standards should be specified exclusively in GLP regulations and not incorporated into testing guidelines.

The emphasis of these position papers was primarily in the toxicology area. However, it became clear as time passed that GLPs needed to be addressed with respect to other studies required for pesticide registration, namely, the chemistry studies. This then brings me to the current topic of this symposium.

NACA GLPs For Chemistry Studies

In the spring of 1985, the Research Directors Committee of NACA formed an ad hoc Subcommittee on Good Laboratory Practices for Chemistry Studies. The Subcommittee was composed of specialists from fifteen member companies who were responsible either for the management of these studies or the quality assurance aspects of these studies. It is important to note that by 1985 a considerable number of companies had already established a quality assurance unit and were well underway with their GLP programs for chemistry studies. The Subcommittee was charged with the development of a document which addressed good laboratory practices standards for residue, metabolism and environmental chemistry studies which were conducted for registration. The document was to be made available to member companies to guide them in the development of their GLP programs. The document was also to serve as a basis for industry discussions with EPA on the subject of GLPs for chemistry studies done for pesticide registration.

A final document was produced in mid-1986. The guidelines were modeled after FDA, EPA and OECD GLP regulations for animal studies. The document addressed residue (laboratory and field), metabolism (plant and animal), and environmental chemistry studies done for FIFRA registration requirements. This paper will not go into detail

concerning the document other than to point out it dealt with the
following subjects:

I.   General Provisions
      A.   Scope
      B.   Definitions
      C.   Applicability to Studies Performed Under Grants and
           Contracts
      D.   Statement of Compliance or Noncompliance
      E.   Inspection of a Testing Facility
      F.   Effects of Noncompliance

II.   Organization and Personnel
      A.   Personnel
      B.   Testing Facility Management
      C.   Study Director
      D.   Quality Assurance Unit

III. Facilities
      A.   General
      B.   Animal Care Facilities
      C.   Animal Supply Facilities
      D.   Facilities for Handling Test and Control Substances
      E.   Facilities for Data Storage and the Collection, Shipping
           and Storage of Samples
      F.   Laboratory Operation Area
      G.   Field Operation Area

IV.   Equipment
      A.   Equipment Design
      B.   Maintenance and Calibration of Equipment

V.   Testing Facilities Operation
      A.   Standard Operating Procedures
      B.   Reagents and Solutions
      C.   Dietary Mixtures of Substances

VI.   Test and Control Substances
      A.   Testing and Control Substance Characterization
      B.   Test and Control Substance Handling
      C.   Dietary Mixtures of Substances

VII. Protocol for and Conduct of a Study
      A.   Protocol
      B.   Conduct of a Study

VIII.   Records and Reports
      A.   Reporting of Study Results
      B.   Storage and Retrieval of Records and Data
      C.   Retention of Records

      These subjects were treated in a general way. The entire
document turned out to be 25 single-spaced, typewritten pages.
During the development of the document, several consultations with

EPA personnel were held to exchange views and seek suggestions and comments on the direction being taken by the Subcommittee. In addition, a one-day workshop was held with company field personnel to develop the section dealing with the field aspects of residue trials.

The preparation of this document had two main benefits. First, it heightened the awareness in the industry of GLPs for chemistry studies - what they are and how to organize to implement them - and it provided a framework for commenting on EPA's proposed GLP regulations. That is, since the issues had been thought through and a consensus reached, evaluation of EPA's proposed guidelines should be facilitated. Other benefits of the activities of this NACA Subcommittee were the members' shared experiences on dealing with GLPs and the exchange of information on EPA inspections. This leads me to the next subject.

EPA and Non-Compliance Audits

EPA started non-compliance GLP audits in about March of 1985. Some companies have experienced up to three such audits since then. Initially there was some misunderstanding with respect to the regulatory aspects of these EPA visitations. Some inspectors were initially under the impression that the EPA's 1983 GLP regulations for non-clinical animal studies applied to the chemistry studies. This was later cleared up and it became understood that these audits were of a non-compliance nature. They (the audits) were to provide guidance to industry on what a GLP program should consist of, and what the EPA would be looking for in a compliance situation. Feedback from our members indicates that they generally found these audits to be constructive. Many good suggestions were received from the EPA inspectors. Hopefully, EPA personnel also learned from these experiences and took away useful suggestions from the companies. I believe the program provided EPA and industry an opportunity to understand differing stances prior to a compliance situation. Hopefully, these have aided EPA in developing regulations. It seems apparent that these audits have prepared industry for regulatory compliance.

There are some points which we believe EPA should bear in mind as we move to a full-scale regulatory situation. The first is with respect to training of auditors. There is a need for consistency. There has been, during this non-compliance phase, some evidence of inconsistency. While this is understandable during this learning phase, we believe it's important to stress to EPA the need for consistency among auditors. None of us can live with a moving target.

The second point is understanding the difference between a GLP audit, a data audit, and a technical audit. It is the industry position that these are three separate entities. All of these are legitimate EPA activities. We realize it's tempting for scientists to delve into the technical details of a particular study when conducting a GLP inspection. However, we believe this is the purview of other EPA activities and the GLP inspector should "stick to the knitting." On the other hand, we realize that one cannot be blind to discrepancies between raw and reported data, and technical

quality issues which may surface during the GLP inspection. Under-
standably, such observations will be reported "up the line," but
they should not be the main focus of the GLP inspector. Those
issues should be left to others to follow-up.

The third area is sample retention. The industry is having
some difficulty with this, particularly with respect to the
retention of crop and tissue samples which have been analyzed for
residues. A "forever" retention criteria creates enormous practical
problems. We would hope that the regulations will provide some
flexibility in this area. A fixed time limit seems reasonable. The
NACA GLP guideline document dealt with sample retention as follows:

"Test system samples which are relatively fragile and differ
markedly in stability and quality during storage shall be retained
only as long as necessary to insure the validity of the study.
There shall be appropriate standard operating procedures for
disposal of test system samples. Samples of test or control
substances, samples of test or control substance diet mixtures,
specially prepared materials, and test system samples shall be
retained only as long as considered valid by the study director."

Proposed EPA GLP Regulations

Specific comments on EPA's proposed regulations are not possible as
they haven't issued as of the writing of this paper. NACA will
study these carefully and submit thoughtful and constructive
comments. There are a couple of points, however, which we would
like to stress. The first is, we suggest it be explicitly stated
that the regulations do not cover efficacy trials. We believe that
these trials present unique differences vis-a-vis laboratory studies
such that the subject be dealt with separately. In addition, we
prefer the terminology Good Field Practices (GFPs) rather than GLPs
for efficacy trials.

The second point is that compliance should be phased in so that
completed or on-going studies are acceptable even if the regulations
aren't precisely met. In addition, there should be some time period
between the publication of the final rule and strict enforcement of
compliance with the regulation. While we recognize the industry has
been well aware that compliance with GLPs was coming, and have had
experiences with GLP audits, the regulations may have some
subtleties which companies have not anticipated. It would,
therefore, take time to "gear-up" to assure that all aspects of the
regulations will be addressed.

Summary

The industry is committed to GLPs. In principle, we support GLP
regulations for chemistry studies. Our views on the specifics must
await the issuance of the proposal by EPA. However, we believe the
industry is prepared to deal with these regulations as a result of
NACA's activities in developing an industry GLP guideline document,
and the experiences gained through EPA's non-compliance audits
during the last two years. While EPA's audit program has been
helpful, compliance with the new regulations should be phased in.

RECEIVED January 29, 1988

Chapter 3

Chemical Aspects of Compliance with Good Laboratory Practices

EPA Perspective on Generic Good Laboratory Practices

Dexter S. Goldman

Laboratory Data Integrity Assurance Division, Office of Compliance Monitoring, U.S. Environmental Protection Agency, Washington, DC 20460

Current Environmental Protection Agency (EPA) Good Laboratory Practice (GLP) regulations under the Federal Insecticide, Fungicide, and Rodenticide Act (FIFRA) apply only to health effects studies. The Toxic Substances Control Act (TSCA) GLPs already include both health effects, ecotox and chemical fate studies. To provide consistency in inspections and enforcement, an extension of the regulations is in development. These are designed as "generic" GLPs, that is, they are intended to be sufficiently broad to cover any test being submitted for regulatory purposes to the EPA without writing new GLP regulations for each new type of study as it becomes accepted by the scientific and regulatory community. Since not all GLP elements apply to all studies the proposed regulations are based upon those principles of GLPs that are applicable to that type of study.

Someplace I seem to remember an old aphorism, maybe from the French, that says: the more things change the more they stay the same. The more things change with GLPs the more nothing changes. I can imagine that about ten years ago there was a great deal of trepidation and confusion about these new regulations that the Food and Drug Administration (FDA) was putting into effect - it would put us all out of business (which it hasn't) - it showed a lack of trust in our basic honesty (which it didn't any more than any other regulation) - it penalized us for the misdeeds of others (the wrongdoers were punished, not everyone) - it dictated who we could hire (it didn't) and so on through a long list of real and perceived ills.

We are discussing today the proposed extension of these regulations and already I am hearing similar comments and arguments, and these from people who should know better. Not long ago I discussed this with someone from a giant corporation, which shall remain name-

less, a corporation whose agrochemicals research section has been
engaged for years in high quality animal testing laboratory work,
both internal and by contract. There is no question in this
company as to where the Quality Assurance Unit fits into the management structure. Suddenly I am told that there are management
questions as to how Quality Assurance principles are to be applied
to testing once we walk outside the confines of the climate-
controlled analytical and animal testing laboratories and get into
field and residue studies.

Basic GLPs as Applied to Analytical Chemistry

I would like to explain the basic principles of both the old and
and the new regulations and try to show how they apply to laboratories conducting analytical chemistry.
    To a certain extent this is a sham for there are many of you
who are doing analytical work related to animal toxicity experiments. Many of you supervise diet preparation technicians. Many
of you either conduct on your own or supervise analytical technicians who conduct the basic work on stability of dosage forms,
on homogeneity of dosage forms, on stability of test chemicals.
You know already that all this work, clearly part of the toxicity
test itself, must be conducted under the relevant sections of the
GLP regulations. So what is there that is different about analytical chemistry related to tests other than these traditional
toxicity tests?
    The answer is nothing.
    For chemists who have been doing work not currently covered by
the regulations I can try to assure you that these regulations will
not work an unbearable hardship on you. In fact, the general consensus is that the higher your level of compliance the more likely
you are to become more cost-effective in your work.
    For those of you who expect me to discuss not principles but
rather details of application of extensions of the GLPs to analytical chemistry, I must apologize for I have little intention of
doing so. There is nothing magical about chemistry when it comes
to compliance with GLPs. The tests are what count and those tests
must be conducted under the appropriate principles of the concept
of GLPs. The contents of this paper would apply to a group of
animal care specialists, or pathologists, or histotechnicians, only
some of the examples will have to be changed.

Generic Regulations and Test-Specific Regulations

I would like to discuss the overall concepts that had a major impact
on the design of the proposed expanded GLP regulations. Previous
regulations have been written around a particular type of test. The
major driving force for the Food and Drug Administration was, of
course, safety tests conducted in vertebrate animals. By adopting
the FDA regulations with a few added items specific to the EPA, the
Agency was locked into GLP regulations directed at short and long
term animal testing. Everything then had to be pulled and stretched
to fit the animal procedures. We could continue this approach and
add GLP regulations for each major type of test. This would be a

never ending task as no one can predict the test procedures that
will be in vogue in 5 to 10 years. Our solution was to go to
"generic" GLPs which said, basically, what I just alluded to,
namely, if you do a test that will be submitted to the Agency,
any test, then it must be conducted under the applicable principles
of good laboratory practices. This is shown in the text sections
on purpose and scope taken from the preamble of the proposed GLP
regulations for FIFRA and TSCA:

PURPOSE (preamble)

"... In addition, EPA is proposing to expand
the scope of the FIFRA GLPs to include the
environmental testing provisions currently
found in the TSCA GLPs. EPA's proposed
revision to the GLPs also extends the scope
of the regulation to include product performance data (efficacy testing) as required
by 40 CFR 158.160..."

SCOPE (preamble)

"..., EPA is proposing to require GLP compliance for all studies submitted to the Agency
which are intended to suport pesticide research
or marketing permits...."

SCOPE (preamble)

"..., EPA believes that GLP standards must
apply whenever data collection occurs. Because much of the test data required by this
Agency are developed in the field, or more
accurately in outdoor laboratories (i.e.,
ground water studies, air monitoring studies,
degradation in soil, etc.), EPA is proposing
to include field testing within the scope of
these regulations...."

I believe that this clearly states the Agency's position on test
compliance.
    The GLP regulations boil down to this: if you submit a study
to a regulatory agency, then this study should have been conducted
in a proper facility by qualified personnel, using properly maintained and calibrated equipment, following written standard
procedures and checked routinely by an independent and qualified
person. All the original data should be archived and it should
be possible to validate the final report of the study by an audit
of raw data.

Changes in Definitions

Once the basic concept had been agreed to then most of the expansion could be accomplished by changing some definitions. For
example, a "laboratory" has become a "test facility" and a "test
facility" can be defined as the place where a test is conducted.
This immediately moves us out of the traditional laboratory and
encompasses field studies, ecotox studies, genetic tox studies,
reentry studies, etc.

The rationale for these changes in definitions is given below
in sections taken from the preamble to the proposed regulations:

SECTION 160.41 GENERAL (Preamble)

"... The studies FDA requires are generally
conducted within the confines of a traditional indoor laboratory. Because the
conditions specified within a protocol can be
artifically manipulated within the traditional
indoor laboratory, the location of these
laboratories is generally not a factor in
determining the quality of a study....
... However, the studies EPA requires are not
necessarily conducted within the confines
of the traditional indoor scientific laboratory... EPA considers any site where testing
is undertaken, for data required by the Agency,
to be a testing facility. The conditions
required by the protocol are not conducive to
artifical manipulation in the field, or other
outdoor testing facilities. Therefore, ensuring
the suitability of the location of these types
of testing facilities is both a valid and necessary
part of EPA's GLP Standards.

The next change in definition has to do with the term that we
we have been using. Please note that I have mentioned field studies,
ecotox studies, genetic tox studies. What is a "study"? The current
GLP regulations define a "study" as shown below:

SECTION 160.3 Definitions (Current)

160.3 (m) "Study" means any in vivo or in vitro experiment in which a test substance is studied
prospectively in a test system under laboratory
conditions to determine or help predict its toxicity, metabolism, or other characteristics in humans
and domestic animals. The term does not include
studies utilizing human subjects or clinical studies
or field trials in animals. The term does not in-

clude basic exploratory studies carried out to
determine whether a test substance has any potential
utility or to determine physical or chemical characteristics of a test substance.

Section 160.3 Definitions (Proposed)

160.3 (m) "Study" means any experiment in
which a test substance is studied in a
test system under laboratory conditions
or in the environment to determine or
help predict its effects, metabolism,
environmental and chemical fate,
persistence, or other characteristics
in humans, other living organisms, or
media. The term does not include basic
exploratory studies carried out to
determine whether a test substance has
any potential utility.

The essential differences between the two definitions are shown
below:

Characteristics of a "Study"

1. What

Current: ...Any in vivo or in vitro experi-
            ment...

Propose: ...Any experiment...

2. Where

Current: ...Under laboratory conditions...

Proposed: ...Under laboratory conditions or
            in the environment...

3. Why

Current: ...To determine or help predict its
            toxicity... In humans and domestic
            animals.

Proposed: ...To determine or help predict its
            effects...in humans, other living
            organisms, or media.

4. <u>But Not</u>

Current: ...Whether a test substance has any
            potential utility or to determine
            physical or chemical characteristics
            of a test substance.

Proposed: ...Whether a test substance has any
            basic utility.

We have now expanded the scope of the regulations to become
consistent with FIFRA's statutory requirements even though this has
meant a departure from the FDA's regulations. Each agency must
meet its own statutory requirements.
We have redefined "test facility", we have redefined "study".
The next definition has to do with the living system that is undergoing the test. Up to now this has been traditionally rodents,
dogs and primates. By using the term "test system" and defining
"test system" as that to which the test substance is applied, we can
now include soil, rodents, primates, bacteria and so on. I will not
go into the specific proposed changes in the text but you will see
the emphasis in the text below which highlights the titles of certain sections:

Section 160.43

    Current: <u>Animal care facilities</u>

    Proposed: <u>Test System care facilities</u>

Section 160.45

    Current: Animal supply facilities

    Proposed: <u>Test system supply facilities</u>

Section 160.90

    Current: <u>Animal Care</u>

    Proposed: <u>Animal and other test system care</u>

"Test system" now includes animals as opposed to the original
text which dealt with animals to the exclusion of other living
organisms and other media such as soil and water. In short,
environmental protection deals with micro and macrocosms other than
those represented by warm blooded vertebrate animals.
    I seriously doubt that the proposed regulations contained any-
thing that is unfamiliar to you. Before I look at some of these
broad principles in a bit more detail, especially as they apply to

the analytical chemistry laboratory, I want to repeat something I
said on the overall principles behind the regulations: if you submit a study to a regulatory agency, then this study should have been
conducted in a proper facility by qualified personnel, using
properly maintained and calibrated equipment, following written
standard procedures and checked routinely by an independent and
qualified person. All the original data should be archived and it
should be possible to validate the final report of the study by an
audit of the raw data.

Major Principles of Good Laboratory Practices

I want now to consider these principles in a bit more detail.

Adequate Facility

First, a proper - or better yet, adequate - facility. This says
nothing about location, construction, utilities, air conditioning,
bench and cabinet color coordination, etc. Adequate from the point
of view that the work can be done properly and safely. Enough room
so that personnel are not getting in each other's way in a
potentially dangerous fashion, enough room to permit the work to be
done properly and safely. Enough room to permit the work to be done
on time, especially if the timing is critical to the outcome, enough
room so that work and eating areas are separated, enough room so
that dangerous materials can be segregated if needed. This is
really a management decision.

Personnel

Second, by qualified personnel. Is the person qualified? That is a
management decision based on job analysis, work and performance
description, etc. No one says that a high school graduate cannot
do a perfectly adequate or better job on some esoteric analytical
equipment than a graduate in chemistry. I am qualified in
analytical chemistry - on paper. I doubt that there is a supervisor who would put me into an analytical laboratory without
extensive retraining and measurable performance criteria.

Maintained and Calibrated Equipment

Third, using properly maintained and calibrated equipment. There is
nothing particularly new about this. We expect to see records of
calibration of equipment, either done as a separate routine or as
part of the analytical sequence. Moreover, we expect to see a
written record of these calibrations and the record should be unique
to that piece of equipment. The equipment should be properly maintained and there is to be a record of this maintenance. You maintain and service it, the dealer does it, the factory does it, whatever. Someplace there is a log that shows when the equipment was
taken out of service, what was wrong, when it was fixed and recali-
brated, and when it was put back into service. Why - because the

regulations say so. Why - because only by examining such records
and comparing them with the dates of the analytical runs can we
gain that degree of confidence so needed for analytical chemistry.
    But, you say, the equipment is self-calibrating and no record
is possible. True and without question, but is it so much to ask
that there be a notebook showing day by day that the self-cali-
brating sequence was run through and everything was hunky-dory?

Standard Operating Procedures

Fourth, following written standard operating procedures. There is
no mystique to SOPs, they are the heart of any test facility. They
assure that everyone follows the same procedure each time, that
there is no oral law that supercedes the written text. How detailed
should these be? There are text books on the market with standard
operating procedures written in - just like you go to the stationary
or office supply shop and buy a standard form will or rental agreement. You have to fill in the blanks. My definition of an SOP is
a written procedure that can be followed by any well informed qualified individual with the complete expectation that the anticipated
result will be obtained. Can an instruction book be an SOP?
Probably not. Most instruction books are written as if
they had been badly translated from a foreign language.
They are frequently difficult to understand. The instruction book can certainly be a part of the SOP, but rarely
the SOP itself.

SOPs and Residue Analyses

Suppose we concentrate for a moment on residue analyses. SOPs are
basic to your operation, to every facet of your operation. You did
not do the field work or the sampling but you assume these were done
properly. You assume responsibility for the samples that arrive on
your doorstep. In this work, chain of custody is critical. Who
receives a box of samples, who logs it in and how, who opens and
inspects the contents, who decides if the storage was correct? How
are the sample numbers logged in, how is the container stored until
the samples are ready for analysis, who assumes custody of each
sample and when? Residue analyses are far more than grabbing a
sample from the freezer, homogenizing it in isooctane and shoving it
into a GC. I repeat, chain of custody and documentation of chain of
custody is critical in this work.

Quality Assurance, Concept and Operation

Fifth, checked routinely by an independent and qualified person.
This is where the concept of Quality Assurance (QA) comes in and
I can assure you, nothing is more important within the concept of
GLPs than QA. As good as you are, the QA Unit has the responsi-
bility of double checking your procedures and your results and

assuring management that the work is being properly conducted and
that there is a high degree of assurance that the numbers can be
relied upon. The QA Unit uses your SOPs - that you have written
and signed off - to check your procedures. The QA Unit is obliged
to sign a statement that is usually the second or third page of a
final report that states that the work was done in compliance with
the GLP regulations and that regular compliance inspections were
carried out during the study lifetime. Absent this statement and
the report will not even be considered by the Agency. The QA
Unit is the most important management tool available to assure you
and the Agency that the report can be relied on. The QA Unit is
that great common denominator in the sky by which we can compare
and contrast facilities and managements.

Extension of Good Laboratory Practices to Field and Residue Studies

I think by now it should be clear that compliance with the GLP
regulations in the conduct of field studies or residue studies is
actually a simple extension of what we have been doing for a decade
in animal studies. The requirements for the residue laboratory
parallel those for the diet analysis laboratory and should present
no serious problems to you. GLPs are a management tool and have
nothing to do with science.
    There are some recurring questions that I might anticipate and
answer now.

Protocols and Reports

Does each study have to have its own protocol? Basically, yes; each
study has to have its own protocol but the protocol can be a canned
protocol in which you just change a few words and refer to the
techniques to be employed since each can be referred to as an SOP
number.
    You must balance your needs for production with your clients'
needs for complete and self-contained reports. Protocols need not
be extensive or elaborate; the required content of a protocol or a
study report is given in the text of the GLP regulations.

Regulatory Schedule

When will the new regulations be final? The proposed rules went
into final Agency internal review on August 4. The statues require
us to give other Agencies - notably the USDA - up to 60 days to
respond in writing prior to publication. Congress has been apprised
of the proposed regulations and has a comment period. Publication
and request for comment should be toward the end of December. This
suggests a review of comments by March 1988 with a final draft
prepared shortly after that. Publication of the final regulations
might occur in May of 1988 with an effective date for TSCA 30 days
later and an effective date for FIFRA - due to statutory
differences - perhaps 90 days later.

Types of Studies Covered

What types of studies will the new regulations cover? Basically any study submitted to the Agency in support of a FIFRA registration or reregistration. There are certain minor exemptions for some physical and chemical characteristics.

Will efficacy studies be covered? Only those efficacy studies required by FIFRA to be reported to the Agency. This will include antimicrobial efficacy, vertebrate pesticide efficacy, etc. It will not include efficacy studies already excluded by Section 158.160 of FIFRA which are part of research and development and ordinarily not called in by the Agency.

Study Director

Must there be a Study Director for each study? Not necessarily. The Department head can be overall Study Director; a senior technician can be overall Study Director. The point is that there has to be someone in overall charge and accountable for the study.

How can I be a Study Director when I had no control over the field operations? That is a good question and I am glad I asked it. That question was not addressed in the proposed regulations and will have to be worked out after the public comment period closes.

Data Recording

Must I record each study in a separate notebook? No. You already have adequate means of carrying multiple studies in one laboratory workbook and you have procedures in place to assure that a client sees only his data during an audit. All data should be recorded in ink and all changes should be authorized by procedures that are already in place.

Compliance Inspections

How often will my laboratory be inspected or studies be audited? The strategy and the policy have not yet been worked out. In the past we have tried to get to a given facility about every two years. I am still unsure as to how we will cover them. We are reasonably familiar with the population of analytical laboratories and are able to keep up with the two-year schedule so far. If you think that we are unaware of your presence you might be right. A laboratory is not put into our inventory until it submits a study or its name comes up as responsible for part of another study. So, even if you are doing studies that you know are coming to the Agency we will not know of your existence until the study is submitted and your name entered into the data base.

Summary

I have wandered afield from ordinary chemistry and that is probably because the principles of the Good Laboratory Practice regulations apply to all scientific disciplines involved in testing of agricultural chemicals for potential toxicity. Those items of particular importance to the chemist will be the calibration and maintenance of equipment, the chain of custody of samples, the proper care of notebooks, the cooperation with the QA Unit and the archiving of raw data.

I hope that this discussion has helped to direct your thoughts and energies toward what will have to be done and, at the same time, assure you that the burden is not extreme.

RECEIVED March 21, 1988

Chapter 4

Directive and Supportive Roles of Management

G. Burnett, J. W. Smith, W. B. Nixon, and P. M. Hernan

Agricultural Division, Quality Assurance Unit, Ciba–Geigy Corporation, Greensboro, NC 27419

> Management has both directive and supportive
> responsibilities for the operations of the quality
> assurance unit to fully achieve compliance with Good
> Laboratory Practice regulations (1,2). The tone for
> the entire testing facility is set by management
> since it is their ultimate responsibility to
> establish and endorse procedures and policies which
> ensure a commitment to quality.

Responsibilities that management must meet for its quality
assurance function can be broadly categorized as directive and
supportive. These responsibilities must be executed by management
by defining and implementing programs, including the establishment
of a quality assurance program, to guarantee that all studies that
management sponsors or conducts are in compliance with Good
Laboratory Practice (GLP) regulations. It is management that
establishes and endorses the concept of quality which is the
cornerstone upon which GLP compliance is built.

In accordance with GLP regulations, management must establish
a quality assurance unit (QAU). Management must decide on the
number of personnel required to provide effective and complete
quality assurance. Further, management must ascertain the
qualifications and training needed for personnel in the unit to
perform effectively. Considerable thought and foresight are
required to structure a QAU which serves as an effective management
tool. The QAU is responsible by being an independent observer, for
monitoring nonclinical studies for GLP compliance and reporting to
management the results of these monitoring activities. Management
must rely upon its QAU to provide judgements whether research is
being conducted according to applicable guidelines and regulations.
Unbiased and accurate information is essential to allow management
to make informed judgements on the quality of the studies conducted
and the overall performance of their testing facility.

Management must position the QAU within the hierarchy of the
organization to vest it with sufficient authority to perform its
management defined functions. Further, in order for the QAU to

0097–6156/88/0369–0024$06.00/0
© 1988 American Chemical Society

maximize its effectiveness, the unit must be positioned separate
from personnel conducting or directing studies in the organizational structure such that the unit can be unbiased in judgements
of the GLP compliancy of the studies audited and facilities
inspected.

After a QAU is established and provided with adequate numbers
of qualified personnel, management has the supportive role of providing training to continually upgrade the skills of the QAU staff.
This is particularly important as management redefines its expec-
tations of the QAU. This should be accomplished by providing
courses, attendance at professional meetings, promoting peer interactions, etc.

Management is responsible for the evaluation of the QAU's
ability to provide effective quality assurance. Since management
relies on the QAU for input on the GLP compliance status of the
testing facility operations and study conduct, and must make
decisions based on this information, management must be confident
of the validity and accuracy of the QAU's findings and recommendations. Thus, management must periodically monitor the actions of
the QAU and conduct reviews of internal QAU SOPs. Management's
assessment program should include a review of QAU personnel records
to determine that the staff is well-qualified, adequately trained
and in sufficient number. Management must also examine the QAU's
adherence to regulatory requirements, ability to adequately defend
company GLP and quality assurance programs and adherence to QAU
SOPs and monitoring schedules. In addition, management needs to
review the completeness and accuracy of QAU records and reports,
and the ability of the unit to effectively interact with all levels
of the organization necessary to accomplish the quality assurance
function. The evaluation program should determine whether the QAU
is performing as desired. Management should develop, during the
QAU assessment, recommendations to improve the overall effectiveness of the QAU operation and schedule programs to implement the
recommendations.

The most important supportive role of management is the
correction of deviations from GLP regulations reported by the QAU.
Management must design, implement, enforce, and, if necessary,
alter existing policies and procedures to prevent recurrence of
reported deviations. Management must act promptly upon the
findings reported by the QAU and ensure adequate responses by study
personnel. Only management has the authority to ensure that
deviations are corrected, or, when necessary, that operational
procedures are changed. This role of management support of the QAU
minimizes the potential for adversarial relations between study
participants and QAU personnel.

Another supportive role that management must accept is the
provision for additional resources when QAU responsibilities are
increased either by management, by revised or new regulations, or
by increased study workloads. Such resources include personnel,
office space, equipment, clerical support, etc. Management must
continually review and balance the allocation of resources to the
quality assurance and scientific study areas to efficiently operate
with the desired level of quality.

GLP regulations further charges management to establish archives for the orderly storage and expedient retrieval of study records, raw data, and/or specimens. Since most organizations choose to structure the archive function under the auspices of the QAU, management not only has the same directive and supportive responsibilities for the archives as it does for the QAU, but has additional roles. Management must provide the necessary facilities for storage of all raw data, study conduct documentation, or specimens under conditions which minimizes deterioration during retention. Archive facilities must have adequate fire protection, the contents properly indexed, and entry limited to authorized personnel only. Management must also identify an individual to be responsible for the archives.

In summary, management must perform both directive and supportive roles in quality assurance programs. Management decisions made regarding the QAU staffing level, qualifications of the QAU staff members, and position of the QAU within the organization provide the basic directives which lead to GLP compliancy. The supportive responsibilities of upgrading the skills of QAU staff members, evaluation of their effectiveness, correcting deficiencies reported by the QAU, and committing the necessary level of resources to quality assurance functions provide essential elements for effective QAU operation.

Literature Cited

1. Fed. Reg. 1978, 43, 59986-60025.
2. Fed. Reg. 1983, 48, 53946-69.

RECEIVED January 29, 1988

Chapter 5

The Human Element of Quality Assurance

Gioya Bennett, Janet Evans, and Norma Roadcap

Department of General Services, Division of Consolidated Laboratory Services, Bureau of Chemistry, Commonwealth of Virginia, Richmond, VA 23219

The human element is a vital factor in all Quality Assurance Programs. It is especially critical in the more individualistic atmosphere of the laboratory as compared to a production line. Chemists must understand and appreciate the need for QA and its application to various laboratory operations. The QA staff must understand the laboratory's functions. A successful program requires cooperative efforts. What is needed is participation, not dictation. The authors, from private and public experience, discuss the need for QA from the chemist's perspective, and describe ways to produce a cooperative and effective program. The goal is the production of valid, supportable data.

The human element is a vital factor contributing to the success of any endeavors undertaken by an organization. Successful implementation of Good Laboratory Practice (GLP) regulations requires recognition of the role of the human element in the laboratory. When confronted with a mandated quality assurance (QA) regulation, such as the GLPs, bench scientists often express concerns regarding the need for such extensive QA practices, the increased paperwork associated with such a program, the time and resource allocation required above and beyond the regular workload, as well as the question of trust. While such concerns are valid, these human apprehensions can best be overcome by involving all levels of personnel in the design, implementation, and evaluation of an overall QA program. GLPs, or any mandated QA policy should be considered the framework around which a comprehensive QA program meeting the organization's needs is developed. While it is the bench chemist who is primarily responsible for the analytical process to which the GLPs apply, the input of the bench chemist is often overlooked in the development and implementation of a quality assurance program. It is desirable, however, to design a quality assurance program which encourages and fosters the interaction of the entire staff.

0097-6156/88/0369-0027$06.00/0
© 1988 American Chemical Society

Such a program is more likely to be well received and generally accepted because it provides a mechanism for bench chemists to express their concerns, and to participate actively in the program.

At the Commonwealth of Virginia's Division of Consolidated Laboratory Services (DCLS), a quality assurance program has been developed which focuses on the interaction between management, the bench analyst, and the quality assurance unit. It is important that these groups in an organization not be pitted against each other as adversaries, but joined together as allies, in a cooperative effort to achieve a common goal. While DCLS does not fall under the jurisdiction of the GLP regulations, the QA program is nevertheless based on a mandated internal QA policy, which contains similar compliance elements. In addition, DCLS does have to comply with requirements similar to the GLPs, such as the EPA regulations for the Safe Drinking Water Act, the accreditation program of the U.S. Dept. of Agriculture's Food Safety and Inspection Service, and analyses conducted under Food and Drug Administration programs and Occupational Safety and Health Act (OSHA) standards.

At DCLS, consideration of the human factor has alleviated negative perceptions, apathy, skepticism and fear toward QA, resulting in a successful program in which all levels of personnel actively participate. It is hoped that the positive experiences of developing the QA program at DCLS will provide some guidance to other organizations faced with the similar task of implementing the GLPs.

Overview of the DCLS QA Program

To understand the approach taken in developing the QA program at DCLS, it is necessary to have a general overview of the organizational structure of the laboratory, and of the program itself. DCLS is the analytical regulatory laboratory for the State of Virginia, and is staffed by nearly 350 employees analyzing over a million samples a year. The laboratory consists of a Quality Assurance Section and the Bureaus of Chemistry, Forensic Science, Microbiological Science, and Technical and Logistic Support. The QA program is based on a Division QA Policy mandated by the Laboratory Director, and a Division QA Plan which establishes broad guidelines for more specific Bureau QA Plans. Each Bureau is divided into diverse analytical sections which operate under even more individualized QA plans that are patterned after the Bureau Plan.

The Quality Assurance and Laboratory Inspection Section at DCLS consists of a staff of five, and is responsible for a number of functions in addition to its QA function. Staff members inspect and certify independent and municipal laboratories in Virginia that conduct analyses covered by EPA's Safe Drinking Water Act. In addition, the section is responsible for administering the safety program for the laboratory. Another function of the section is to evaluate a number of products for compliance with bid specifications in support of procurement activities for the Division of Purchases and Supply. Because of these multiple responsibilities, and the large size of the laboratory, an organizational support structure is necessary to assist this section in the administration of its QA function. The Division QA Policy and QA Plan specify the creation of organizational subunits responsible for QA. These include Division and Bureau QA Teams composed of bench scientists. Individual Bureau QA Plans assign additional QA responsibility to Section Representatives and QA Audit Teams, which again involve bench analysts.

It should be noted that the basic elements of the DCLS QA Program closely parallel those set forth in the GLP regulations. Each of the major components of the GLPs is addressed in the BOC QA Plan.

Involvement of Personnel in the Development of the QA Program

The DCLS QA Program evolved as a result of participation and input from all levels of personnel from the outset. The QA program in the Bureau of Chemistry was the first to be developed at DCLS, and serves as the prototype for QA programs in the other Bureaus, although each Bureau is expected to address its own unique functions in the preparation of a QA plan.

The program was mandated initially not only by a management policy, but also by a Strategic Plan for the laboratory. Both documents emphasized management support for the development of a QA program, as well as a commitment to involving the entire staff in this process. The degree of management support for the program is reflected in the policy statement, " Only safe working conditions for all personnel have higher Division priority" (2). An overall Division QA Plan was written by management in consultation with the QA Section. Management immediately involved bench chemists in the program development phase by assigning analysts to serve as Bureau QA Team Leaders. In the BOC, the team leader then chose two additional team members to assist in preparing a Bureau QA Plan. These individuals were selected because of their technical expertise and knowledge of QA principles. They accepted this additional responsibility because of their commitment to the concept of QA, as well as their desire to have input into the program. This core group of people dedicated to enhancing the quality of work performed was viewed as an integral part of the laboratory function. The joint efforts of these individuals over a period of four months resulted in a Bureau QA Plan that was reviewed and accepted by management and the QA Section as a guideline for the preparation of more specific Section QA Plans. Throughout this process, management and the QA Section provided guidance and support, but bench scientists serving on the QA Team were free to develop the QA program within the framework of the mandated policy.

After the development of the Division and Bureau QA Plans, involvement of the entire staff was sought. Section Representatives were selected to oversee the writing of individualized Section QA Plans. Work assignments were made each week, and assignments from the previous week were reviewed, to ensure timely completion of this task. It was at this stage that problems were first encountered with resistance and negativism toward the

concept of a mandated QA program. The program was initially
perceived as a major change in laboratory practices, and analysts
were concerned about the added demands it would place on existing
resources. Because the QA Team itself was composed of bench
analysts, it could effectively understand and empathize with these
feelings. It was necessary to reassure analysts that their
current laboratory performance was good, but that documentation of
quality work is essential to maintaining laboratory credibility.
It was pointed out that most of the proposed QA practices were
already in existence, but just needed to be formalized. Management's willingness to allocate time and resources to the implementation of the QA program alleviated concerns over the consequences
of reduced sample output while instituting additional QA practices. As understanding of the QA program grew, each laboratory
within the BOC learned to develop an atmosphere of cooperation and
accord, whereby everyone strove to meet their common QA goals.
The involvement of bench chemists with well thought out protocols
resulted in functional QA Plans for each section in the Bureau.
The importance of this work was felt by all levels of personnel,
from technicians to the upper level of management.

During the process of writing the Section Plans, training and
education were other essential elements which influenced the bench
chemists in overcoming their apathy and skepticism. At this time,
the representatives were trained on a routine schedule by an
enthusiastic trainer (member of QA Team) on how to adhere to and
understand the principles presented in the Bureau QA Plan. When
Section Plans were completed, representatives shared knowledge
with their individual lab sections through organized QA meetings.
These training meetings initiated the process of familiarizing
everyone on their QA guidelines and addressed each element of the
program in detail. This internal training again involved all
levels of personnel, and focused on such topics as: development of
a QA program, statistics and control-charting, sampling procedures, conducting audits, etc. Slide programs and video cas-
settes were also utilized for the in-house training. The U. S.
Department of Health and Human Services of the Food and Drug
Administration publishes a Catalog of Courses and Training
Materials which can be a valuable resource for such training.
Benefits of QA education were noticed quickly, because the program
did not seem as overwhelming when taken a step-at-a-time.

DCLS has also taken advantage of external training in QA. Management has been supportive in sending supervisors and chemists to
courses provided by the Association of Official Analytical
Chemists (AOAC). After attending such courses, some initially
"less than enthusiastic" personnel actually returned with innova-
tive ideas and contributions. The AOAC, other organizations, and
of course ACS present excellent opportunities for QA short courses
and seminars. Since education leads to understanding and acceptance, it can be the best public relations campaign for a QA
program.

With the completion of the QA Manuals and training of personnel in QA practices, the laboratory proceeded into the implementation phase of the program. At DCLS, audits are conducted internally at the Section and Bureau levels, and externally by

regulatory agencies such as the Environmental Protection Agency (EPA), the National Institute for Occupational Safety and Health (NIOSH), the United States Department of Agriculture (USDA), and the Nuclear Regulatory Commission (NRC) where appropriate. The BOC QA Plan specifies that each section be audited at least annually by a BOC QA Audit Team. In addition more frequent audits are conducted internally by the individual sections. Initial audits were conducted in each section of the BOC to assess the status of compliance with the program. Audits were approached as a means for aiding the growth and development of a section, and as a positive learning experience, rather than reflecting negatively on a section. The BOC Audit Team is composed of a member of the QA Section, who serves as the Audit Team Leader, and a BOC QA Team member from a laboratory other than the one being audited. The appropriate Section Representative also serves on the Audit Team in an advisory capacity to provide information on the section's QA program. When lab personnel become part of the audit process, they are usually more willing to accept and learn from the audit findings. In addition, the effectiveness of the Audit Team is enhanced by the combined efforts of someone knowledgeable and skilled in QA practices, as well as someone with technical expertise. Participation of bench scientists in the audit process affords them the opportunity to review laboratory operations and techniques from a QA perspective, and to understand the importance of these activities. An added benefit of the involvement of analysts on the Audit Team has been an increased understanding of and respect for the work performed in other sections.

Members of the Audit Team attempt to find a way to accomplish their goals with as little disruption and as much accord as possible. During audits, the Audit Team tries to focus not only on weak areas, but looks for accomplishments and beneficial situations. Audit findings are summarized in a formal written report. Before this report is submitted to management, however, it is presented to the bench scientists in the form of an oral debriefing. Personnel are encouraged to respond to the audit results in writing. Feedback from the bench scientists is considered a valuable aspect of the audit proceedings. When disagreements over audit results occur, management actively fulfills its responsibility by resolving these conflicts. The importance of the interaction of all levels of personnel is therefore demonstrated in the audit process, as it is in all phases of the QA program at DCLS.

The active participation of the entire staff in a QA program is essential. Responsibility for QA is therefore included in every employee's position description in the BOC. This emphasizes the importance of QA in the laboratory, and demonstrates an expectation of participation in the program. Individual performance standards addressing QA provide a mechanism for evaluating an employee's participation in the program, and rewards those demonstrating a positive attitude towards QA. This also helps establish positive role models for participation in the program.

Currently the DCLS-BOC QA Program is entering into its second audit cycle. The Bureau and Section QA Plans are undergoing an annual review and update, with input from the entire staff.

New Section Representatives have been appointed to serve on the
Bureau QA Team. In this way bench scientists will have an opportunity to rotate into a more active role in the QA program. It is
hoped that eventually most analysts will have served in this
capacity. The DCLS QA Program is considered to be an active,
dynamic process, undergoing review and change as needed, based on
the support and input of all laboratory personnel.

Industrial Applications

One of the managers at DCLS was involved with federal military
contracts for major development programs in a previous position
with Hercules, Inc. Similar broad based participation was used to
develop specifications, analytical procedures, and quality control. Resolution of concerns over differences between the company
and the responsible federal agency was accomplished through
negotiations on a level of mutual professional respect. In many
cases, the result was improved procedures in' terms of effectiveness and economy.

   Implementation of the GLPs at the Analytical Laboratories of
the DOW Chemical Co. also followed a somewhat similar approach as
that taken at DCLS. Through meetings and the publication of the
DOW Analytical Laboratory Practices, laboratory personnel were
familiarized with the regulations. The initiation of an audit
system involving all groups of laboratory personnel assessed the
degree of compliance with the GLPs, and identified areas needing
additional clarification or attention. This approach resulted in
a growing awareness, acceptance, and compliance with the GLPs, and
is another example of the successful involvement of lab personnel
in a QA program (5).

   The experiences at Hercules and DOW demonstrate that a team
work approach to QA can be successful in an industrial as well as
a government laboratory setting.

The Importance of Communication

The importance of communication was evident throughout the
development and implementation of the QA program at DCLS. A communication network between the QA unit and all levels of management and laboratory personnel is essential to the success of such
a program. At DCLS, this communication network consists of regular Section, Bureau, and Division QA Team meetings, and quarterly
reports from the individual Sections and the Bureaus to the QA
Section and management, in addition to audit reports.

   Employee interactions on QA matters at DCLS were enhanced by
utilization of such communication skills as listening attentively,
discussing rather than arguing, showing empathy and sensitivity,
and encouraging participation. Persuasiveness and influencing
skills were important to the QA section in convincing management
and laboratory personnel of the need for QA policies and changes.
These same skills aided management and bench analysts in interact-
ing with the QA Section in presenting their needs and capabilities. Knowledge reflecting real understanding, sincerity showing
full support and belief, empathy demonstrating respect for other

attitudes and beliefs, and enthusiasm generating participation are persuasive qualities that were effectively employed to achieve a desired outcome.

Semantics was found to be another important consideration in QA communications, because the language used helps to achieve the desired effect on participants. Using positive sounding words rather than negative comments to describe QA operations elicits good feelings and stimulates participation in the QA program. In the Air Force, for example, QA audits or inspections are referred to as "staff assistance visits". In this case, use of the term "assistance" reflects an intention of helping and working together on problem areas, whereas the words "inspection" or "audit" have a more negative connotation. Similarly, the title QA "Officer" places emphasis on the policing function of this position, while QA "Supervisor", "Director", "Coordinator", or "Specialist" describe a more positive image. Audit and QA reports should also be carefully worded, clearly describing corrective actions needed while emphasizing positive findings.

At DCLS, QA personnel, management, and lab personnel work together to solve QA problems and to take corrective actions. Effective communication throughout the problem solving process is essential. Resolution of a QA problem follows these basic inter-active steps: defining the problem accurately, generating possible solutions, evaluating solutions, deciding on a mutually acceptable solution, implementing the solution, and evaluating the solution. This problem solving approach again recognizes the importance of the interaction of all members of the organization.

Summary

The human element is the single most valuable resource in a QA program. The DCLS QA Program is designed to optimize the interaction and involvement of all levels of personnel. In recognition of this fact, management sets the policy and provides support and resources for the program; the Quality Assurance Section monitors the program, conducts audits, and provides guidance and support to the Bureau QA Teams; the Bureau QA Team acts as a liaison between the bench scientists, and the QA Section and management, oversees the overall QA program for the Bureau, participates in the audit process, and provides guidance to the sections and individual analysts; and bench scientists provide input into the QA program, conduct their work in accordance with the QA program, serve on the Bureau QA Team, and participate in QA audits. Although the specific elements of a QA program may be different for each organization, the involvement of personnel in the creative development of such a program can be effective in any laboratory setting. The recognition that bench scientists are the mainstream of the QA program contributed to a growing awareness and accep-tance of the mandated policy at DCLS. The outcome of this effort has been the demonstration of laboratory credibility and a high degree of professional integrity.

Acknowledgments

The authors gratefully acknowledge the input and support of
Dr. Albert W. Tiedemann, Jr., Director of the Virginia Division of
Consolidated Laboratory Services, (DCLS), Mr. Edward E. LeFebvre,
Director of the Bureau of Chemistry, and Mr. Warner Braxton,
Quality Assurance and Laboratory Inspection Section Supervisor, in
the preparation of this paper. We also thank Ms. Maureen Barge,
Quality Assurance Specialist, FMC Corporation, and
Dr. Paul Lepore, of the Food and Drug Administration's Office of
Regulatory Affairs for their encouragement and review of this
work, and Ms. Helen Shires for her clerical support. Most impor-
tantly, we recognize the contributions of our co-workers and
fellow employees at DCLS to the success of our QA Program.

Literature Cited

1. Code of Federal Regulations, "Protection of the Environment",
   40, (160), July 1, 1986.
2. Commonwealth of Virginia, Department of General Services,
   Division of Consolidated Laboratory Services Policy A-16,
   Quality Assurance Program, April 1, 1985.
3. Commonwealth of Virginia Quality Assurance Plan for Bureau of
   Chemistry, July 1986.
4. Commonwealth of Virginia Quality Assurance Plan for Division
   of Consolidated Laboratory Services, April 1985.
5. Koch, M. V., Bulletin of the Analytical Laboratory Managers
   Association, Special Edition 87-1, 8-18.

RECEIVED January 29, 1988

Chapter 6

Integration of Quality Assurance into Analytical Laboratories

Joseph B. Townsend

Bio/dynamics, Mettlers Road, East Millstone, NJ 08875

Integration of Quality Assurance concepts into the
laboratory is the key to GLP compliance and is usually
accomplished in three phrases under the guidance of the
QAU. In the Management Phase the basic plan is formulated based on policies decided upon by management. In
the second phase the QAU prepares the laboratory for the
final implementation phase. General rules for integration are given as are levels of acceptance that may be
expected from laboratory personnel.

Experience has shown that the proper integration of Quality
Assurance concepts into the laboratory is the key to compliance.
This presentation includes some suggestions for proper
integration and for easing the analytical chemistry laboratory and
more appropriately the analytical chemist into the new world of the
Good Laboratory Practice Regulations (GLPs). It would not be
appropriate to tell you that all suggestions included herein are
tried and true and that by following them, the course you take to
compliance will be smooth and uneventful. Let it simply be said
that the purpose of this paper is to relay to you some of the things
that the author did right and some of the things that in hindsight
should have been done. Although the remarks that follow may often
appear to be directed to persons who are facing integration for the
first time, they are intended also for those persons who have passed,
or who are passing through the experience now. These remarks should
be pertinent to the bench chemist who after all is the key to
compliance, as well as to the extraordinary man or women who is given
the responsibility for integration.

Although compliance is largely a human problem with all of the
vagaries attendant thereto, it is a matter of common sense and can
be addressed in a coherent logical manner. To simplify the
description of the process, it has presumptuously been broken into
three phases. These phases obviously are not discrete and do
overlap, but they should help illustrate the several points that are
to be made.

0097–6156/88/0369–0035$06.00/0
© 1988 American Chemical Society

Management Phase

Management must take an active role early in the integration process, because there are important high level decisions that must be made before implementation can proceed further. If there is presently no Quality Assurance Unit (QAU), the first decision should be to assign overall responsibility for establishment of the QAU, and therefore for GLP integration, to a person competent to carry the task to completion. It is important that in addition to being innately capable, this person must be given full authority by Management. He or she must be given the time, have some knowledge of analytical chemistry procedures, have a desire to learn the GLPs, be of unlimited patience, be articulate, be persuasive and have the fortitude to pursue the job to its conclusion. The selection of the proper person to orchestrate this project could be the single most important step to compliance. The person designated as responsible for integration should begin by mentally walking through the process and preparing an integration plan - at this point the plan should be simple and flexible, to be modified as the process continues.

Very early, the Plan will indicate that he or she must return to Management for a number of obvious policy decisions. Depending on the persons who must be involved, securing approval could be tedious, so it may be expedient to present proposals for management approval, rather than to ask for complex policy decisions directly. It should also be suggested that questions from people in management about the process, be resolved early and as integration progresses, that they be kept informed.

Some questions that may have to be answered by management are;

- If a Quality Assurance Unit is not now in place, how will it be organized, and to whom should it report?
- Should the entire analytical laboratory be made to follow the GLPs when only just a fraction of the work will require it?
- Should laboratory modus operandi be changed, such as abandoning the use of notebooks and going to discrete data sheets?
- Are there facilities for storage and separation as required by the GLPs, or will they have to be constructed?
- Who should handle inspections by the agencies and what is the company inspection policy?
- Who will be responsible for training, and what will be the training policy?
- Who, for GLP purposes, constitutes management? Who can be a Study Director?
- Who is to prepare SOPs and how should they be organized? By operating group or functionally? What should be the mechanism for change, distribution and authorization of SOPs? Should equipment SOPs be limited or should they include items such as stirrers, hand calculators, etc.? How should the historical SOP file be handled, and by whom?

Preparation Phase

The second Phase, the Preparation Phase, is the phase in which the members of the Quality Assurance Unit and the laboratory personnel must collaborate fully. It will require the most effort and will for the first time bring many employees to the realization of what compliance with the GLPs will require of them personally. This step should be characterized as one of training as well as one of preparation. Training sessions, seminars and informative discussions with professionals and non-professionals, conducted with the intention of dispelling "GLP antagonism" and misinformation (and there will be plenty of that) are imperative. "GLP shock" will be ameliorated if through attendance at meetings outside of their company or institution, employees have an opportunity to discover that other people in other laboratories are experiencing the same problems they are experiencing. As integration progresses, it is very important that employees be kept informed about what is happening.

Meetings with employees have the added advantage of identifying persons in the laboratory who can be counted on to support and lead the way and those who cannot. Persons who have strong negative feelings about the GLPs often reveal their feelings in these meetings and can be given special support.

In addition to keeping people informed, an equally important rule to be exercised in this phase is to involve them in the integration process. For example, allow the operating groups to write their own SOPs, or involve them in preparing lesson plans. Imposition of rules without at least review and comment by the people who will be required to follow them, could present problems. By involving laboratory people, both technicians and professionals, the greatest opportunity to encourage compliance is presented. Some of the tasks that must be attended to in this phase are:

- Writing and review of SOPs including, maintenance of equipment, quality assurance, report preparation, and proper preparation and handling of protocols.
- Implementation of formal training procedures, including lesson plans, educational aids and documentation of course work for each employee.
- Preparation and update of job descriptions and curricula vitae.
- Establishment of proper archives.
- Establishment of a program for proper labeling of reagents.
- Establishment of regulatory inspection procedures.
- Validation of computer systems.

In this phase, the person in charge of Quality Assurance who is responsible for integration should be prepared to encounter procedural questions from laboratory personnel. The answers to these questions are in many cases judgement calls and may, when taken together, set policy, policy that may in the future take prodigious amounts of effort to change once the laboratory has gotten comfortable with them. When answering procedural questions, the Quality Assurance Manager must therefore be careful with his/her answers and often consult with others inside and outside of the company before being definitive. They should document their answers and be consistent, make a policy file or book for reference, and make it available to all. Fortunately or unfortunately this process draws the QAU Manager into the position of being the GLP interpreter, the company "Expert" on Good Laboratory Practice matters. This position is not to be considered lightly since he or she may be called upon later to defend decisions that are questioned by a regulatory inspector.

Implementation Phase

Finally, in phase three, one finds that the laboratory has passed out of the Preparation Phase and is now in the Implementation Phase, which is the longest and in some ways the most difficult. It is during this period that the laboratory begins to operate under the new rules and it is when all members of the analytical team, regardless of their opinion of the GLPs, must come into conformance. It is the period of testing and modifying the processes everyone worked so hard to prepare.

Initially there is some confusion as some people suddenly realize that GLP implementation is about to become a reality and that they must learn "what this is all about." In addition it may happen that one person (hopefully no more) will "actively" or covertly resist. The policy regarding these persons should be to be consistently firm but not confrontational. You should be willing to lose many battles as long as you win the war. It should be stressed that if people have been kept informed and if the Preparation Phase was adequate, most people will cooperate and assist in bringing the laboratory into compliance. The "resistors" are a small minority whose influence wanes and who eventually will become cooperative. "Resistors" eventually become clear and surprisingly, vocal, supporters of the GLPs.

People appear to progress through several stages before they finally reach complete acceptance. While some people progress faster than others, they did not in our sample appear to sort themselves out by age, or sex, or in hindsight by any other characteristic. This phase, however, will try the patience and perseverence of all members of the Quality Assurance Unit.

Listed below are phrases To describe each level of acceptance. They show characteristics that may be typical, as acceptance advances from a period of resistance and questioning to an attitude of active support and compliance.

A) Resistance:
Exhibited in some persons by mild to severe antagonism, defensiveness, sometimes anger. Because of the visibility and the role of Quality Assurance, focus of resistance is often placed upon Quality Assurance. Extensive questioning of inspection findings by some, who take inspection findings personally. Some complaints to QA monitors about unfairness. Extensive explanatory responses to QA findings. A feeling by some that GLPs restrict science. Comments are heard like "These rules shouldn't apply to us" or "I spend all my time on paperwork - the GLPs will kill science."

B) Resignation:
While some persons may respond to the new rules by trying to ignore them, complaints to management begin to appear indicating that the GLPs are restrictive and are arbitrary and capricious. The specifics of inspection findings are properly addressed but not the principles they illustrate. Sometimes cryptic responses

to inspection findings are noted, and some suggest that the
company, or they individually, fight back by showing the FDA or
the EPA the ludicrous nature of the GLPs.

C) Acceptance:
   Requests are made for copies of GLPs. QA monitors begin to be
   asked questions and it is requested that meetings be held to
   explain the GLPs. Responses to inspection findings show more
   acceptance with less antagonism, and there is less personalization
   of QA findings.

D) Support:
   QA findings begin to diminish in number, and fewer "significant"
   inspection findings are noted. Discussions of GLP interpretations
   and fine points are elicited, and professionals and technicians
   begin to take pride in their level of compliance. Employees begin
   to show understanding of GLP concepts in their responses to QA
   audit reports, and suggestions are made spontaneously for
   procedural modifications and changes. Professionals and
   non-professionals begin to knowledgeably and rationally challenge
   inspection findings of QA and regulatory Inspectors.

   To achieve the point where people begin to support GLP concepts
should be a source of great pride to the people in the laboratory as
well as to the persons in Quality Assurance responsible for
integration, and it should be a source of comfort to management. But
integration efforts can't stop there. Compliance is relative and it
is continous. As new people are employed, as the regulations change
and as internal business and organizational changes take place,
changes must take place also in the way your company addresses the
GLPs.

Regulatory Inspections

At this point, a comment should be made about preparing for regulatory inspections which should be a major consideration when
integrating GLP regulations into the laboratory. Regulatory interface is a consideration that is often forgotten. Although the
Quality Assurance Unit Manager does not have to have the task of
hosting regulatory inspection teams in the laboratory, a good QAU
Manager is, because of his/her intimate knowledge of the GLPs,
company policy and regulatory interpretations, an ideal candidate.
The person given this charge by management who ever they might be,
should set to the task of preparing standard operating procedures
and conducting training sessions in anticipation of an inspection.
   Because some regulatory inspectors lack training and experience
in the laboratory and because some unfortunately do not have a clear
understanding of their charge, it is imperative that the person from
your laboratory that is designated as their host, be knowledgeable
and have a clear understanding of regulatory constraints as well as
the GLPs.
   It should be noted that this paper has focused on GLP
modifications proposed by the EPA for the analytical laboratory, it
also should be noted that similar GLP regulations may soon follow
from other agencies such as from agencies outside the country.

Principles of Compliance

There are several principles that are key to compliance with the
GLPs. These principles may appear self-evident but they need
frequent reiteration. In a busy laboratory it is often difficult
to remember an apparent abstraction when it may not directly apply
to the task of the moment. These points should be reinforced during
training sessions.
- Proper documentation is not discussed specifically in the
  regulations, but is one of the most important precepts of the GLPs
  and is one of the hardest for people to practice.
- The creation of strong audit trails is also not mentioned in the
  Regulations and is a precept often not properly adhered to.
- SOPs should be written to reflect what is now being done, not what
  will be done someday.
- The definition of "raw data" should be clearly understood. It may
  include items that on gross inspection may appear strange, but raw
  data are the product of your efforts and must be handled in
  conformance with the GLPs.
- The role of the QAU should be clearly understood. The QAU is an
  observer and reporter. It is not a judge or a policeman and it
  should not pass judgement on the scientific aspects of studies.
- Do not mistake Quality Assurance for Quality Control.
- GLP compliance is the responsibility of the Study Director and not
  Quality Assurance.
- Quality Assurance is not a safety net and it does not purify or
  sanctify. Don't assume QA will pick up all errors.
- Because of the nature of the regulations, interpretations are
  required. Be reasonable, but in lieu of interpretations from the
  Agency, interpret in favor of the laboratory and be prepared to
  defend.
- Do not solicit interpretations from a regulatory agency unless
  you are prepared to live with the answer.
- Blind, unreasoned compliance with the GLPs is sure trouble.
    Those who have had experience in the laboratory of adjusting to
the GLPs have recognized at least some of these comments as being
familiar.

Summary

For those who will be required to face the experience of integrating
the GLPs into the laboratory for the first time, several points
critical to the process should be mentioned in summary.
- Assign one carefully chosen person to guide the transition. The
  best choice is the Head of the Quality Assurance Unit.
- Prepare a plan and obtain management commitment for policy.
- Conduct GLP training for all laboratory employees, professional and
  non-professional, and keep them informed about integration
  progress.
- With direction from the Quality Assurance Unit, have laboratory
  personnel themselves prepare for implementation. Do not impose
  changes from the top down.
- Someone will be required to interpret the regulations - carefully
  consider and record interpretations for their future impact on the
  laboratory and their acceptance by the regulatory authorities.

RECEIVED January 29, 1988

Chapter 7

Good Laboratory Practices and the Myth of Quality

Maureen S. Barge

FMC Corporation, Princeton, NJ 08543

Good Laboratory Practice Standards are intended to assure the quality and integrity of data submitted in support of pesticide registration. Recently a debate has arisen around how quality is defined in this context. GLPS provide a system for the reconstruction of a study through a paper trail. While GLPS are designed to be sufficiently flexible so they can be adapted to a variety of studies, they do not define specific measurements of quality. Although many people can recognize quality work, few can readily define the parameters used to measure quality. Therefore, by whose definition can/should quality be defined?

Quality can be defined as the characteristics or attitudes associated with excellance or superiority. Therefore to develop a good laboratory practices program which supports quality, one merely needs to develop or write down those characteristics or attitudes that reflect a superior operation. Start by asking someone who manages a quality operation to put some concepts down on paper. This has to be easy, doesn't it, because everyone knows what quality is. Easy that is until one begins to implement a good laboratory practices program in a chemistry laboratory. Here the enigma begins, because quality is a highly subjective personal value and because of this, the existence of good laboratory practice standards (GLPS) alone can not guarantee that the reported work is scientifically sound. Unless the program addresses science and good record keeping collectively all GLPs will do is ensure that the documentation was done in the lab, not the quality of the work.

0097–6156/88/0369–0041$06.00/0
© 1988 American Chemical Society

Good Laboratory Practice Standards, as prescribed by
40CFR160, are "intended to assure the quality and
integrity" of studies submitted in support of pesticide
registration. But, what is meant by quality in this
context? The GLPS provide a system for the
reconstruction of a study through a paper trail
documenting everything from the qualifications of the
personnel conducting the study to what raw data and
records are to be retained and how they should be
archived. What the GLPS do not do, in point of fact, is
define the parameters used to measure quality. GLPS,
however, do allow a reviewer to recontruct the study
through the paper trail which will make the quality or
lack of quality quite obvious.

Quality is a very relative concept in that the
control measurements in the form of checks and balances
differ from lab to lab and chemist to chemist. Each
chemist through education, training and experience has
a preconceived idea of what controls are necessary to
produce good science. The inherent variety of
backgrounds and personalities which come together in the
work place produces the same variety in the concept of
quality. There are many resources available providing
guidance on how to conduct and document a study. The
EPA has published Standard Evaluating Procedures,
Pesticide Assessment Guidelines, and the Data Reporting
Guidelines. In addition, scientific journals and
proceedings from symposia, such as this one, help
chemists to determine what is currently being done in
other laboratories. The minimum level of the quality of
the science is dictated by the management of each
company. Quite frequently the chemists themselves
however, will set the quality control measures to be
adhered to through the use of well written standard
operating procedures (SOPS) at an even higher level of
performance standards than management.

Historically, the original GLPS were designed
primarily for toxicology studies. As a result, those of
us who have been conducting theoretically, non-regulated
studies, such as residue and metabolism, have
experienced a great deal of frustration in our attempts
at compliance. The methodology, terminology, and logic
which exist in a toxicology study may not prevail in a
chemistry study. In trying to fit this square peg in a
round hole, we have in the past few years ended up in
the regulatory limbo called "the spirit of compliance".
Now, however, the game has changed and the 'spirit' of
compliance is no longer good enough. The most recent
revisions to the GLPS are an effort to design one set of
regulations for all studies, that is, generic GLPS.
Unfortunately to design generic GLPS means that the
final document must be filled with generalities. Thus a
document stating what is expected to be accomplished,
but no measures as to how to do it. Or in other words,

what our goal is, but no game plan. In essence the GLPS provide requirements for the thorough documentation needed to recontruct a study, but the soundness and quality of the science remain to be determined.

Good Laboratory Practice Standards

The GLPS is a very thorough document encompassing all facets of a study including the organization and the people the facilities, the equipment, the protocol and conduct of the study, the records and the reports with requirements for documenting a study from beginning to end. It is important to recognize the fact that the GLPS require documentation, but do not provide the standards of performance. A key point that is readily overlooked.

For example, Subpart B Organization and Personnel states that each individual shall have a combination of education, training, and experience necessary to perform the assigned functions. In addition, the testing facility is required to document this training and experience. Here the GLPS do not set the criteria to equate a particular level of education to match a particular task. It does however require management to maintain records to justify the matching of people and their job functions. These records are one of the first items requested by the Agency during an audit, therefore the capabilities of the personnel conducting the studies will become quite evident.

By utilizing such terms as suitable, sufficient, adequate, and appropriate, the scientists who are responsible for writing the GLPS have acknowledged that the responsibility for scientific judgement must remain with the chemists who are conducting the study. Subsection 160.63 of 40CFR160 requires equipment to be adequately inspected, cleaned, maintained, tested, calibrated, and/or standardized. With the large variety of instrumentation used to conduct all the various types of studies which are regulated by the GLPS, it is not feasible for the measurements used to determine adequate maintainance and calibration to be listed in the GLPS. However paragraph (c) of this subsection does require written records which are usually in the form of a logbook for each instrument containing the necessary information to support proper use and care. Here again, through a review of the records a decision can be made on the quality of the science.

Another similar example is the reference to Standard Operating Procedures. The testing facility shall have written SOPS that adequately ensure the quality and integrity of the data generated. This is an opportunity for the chemist to ensure that the quality of the science is maintained uniformly and consistently throughout the laboratory, no matter who is doing the

work. Well written SOPS can set the standards of
performance necessary to maintain the level of
excellance desired by management and dictated by the
science. However it should be expected that SOPS will
differ from company to company or university to
university. The bottom line is there are no rules on
how to write SOPS nor should there be. It is the
scientist conducting the study who should use good
judgement in writing the SOPs to build in the quality.
Again an Agency auditor will want to review the SOP
manual to determine the quality which has been built
into both the laboratory and quality assurance standard
operating procedures.

In addition, there are no criteria in 40CFR160 for
the resolution of the chromatography. What numerical
value should be placed on a sample found to be non-
detectable? What should the expiration date be on an
analytical standard or a bottle of methylene chloride?
You won't find the answers to these questions in the
GLPS. If these answers cannot be found in the GLPS,
where can they be found? The responsibility for the
limiting factors needed to produce quality work begins
first with those conducting the study.

Quality Control vs Quality Assurance

When speaking of quality, it is necessary to make a
clear distinction between the two components which
develop a quality program - quality control and quality
assurance.

Quality Control refers to the tools a chemist uses
to measure the accuracy and precision of the methods and
procedures.

Quality Assurance is the system of monitoring,
inspecting, and auditing which assures that the work is
documented and conducted according to protocol and the
laboratories standard operating procedures from the
conception of a study to the review of the final report.

While these are two separate and distinct
activities, each must complement the other to ensure a
quality program. Day to day quality control in the
laboratory is the obligation of the chemist. The
chemist develops the methods, calibrates the
instruments, and with management approval develops the
standard operating procedures for the laboratory.
Quality control is running duplicate samples, reagent
blanks, fortification samples, linearity checks and
confirmatory analyses.

Quality Assurance, the responsibility of the quality
assurance unit, is the nitpicking, but totally
necessary, job of determining the quality of conformance
to regulations established by managers and their
chemists and is done via audits and inspections. To
develop a thorough quality program both quality control

and quality assurance measures must exist. Quality science is of limited value without the supportive documentation mandated by the GLPS and monitored by a Quality Assurance System. Quality documentation without quality science may prove the study to be invalid and it is certainly not the goal of any laboratory to produce invalid data. However, we are now in an age where quality science without documentation will also result in an invalid study.

Responsiblity

The responsibility for standards of performance of both the laboratory and the quality assurance unit must lie first with management. A quality program must be a triad composed of management, chemists, and quality assurance. Management must provide the necessary resources, (both human and material), and the necessary time to not only do the work but document it. The chemists develop valid procedures for conducting studies which management approves and then the quality assurance unit ensures the study director and management that each study is being conducted according to protocols and standard operating procedures. Perhaps with this triad in mind, it will be easier to understand what each others jobs entail and that everyone is responsible for the quality of the study.

Conclusion

GLPS may or may not add quality to a laboratory, hence the title of my paper, "GLPS and Myth of Quality". For those laboratories producing quality science but poor documentation, GLP compliance will force the chemist to think about the importance of his research in supporting registration. For those with poor science, it will be easier to detect the poor quality and force these chemists to develop better programs. While it is the responsiblity of EPA to ensure a relatively safe environment to the American public, it is our responsbility to produce good science and document it. We can not go to the IRS and say 'Well, I calculated that I should have paid $1000 in taxes this year' without supportive documentation. Well neither should you expect the EPA to grant a registration for a product by merely stating that the compound is not a hazard, has no detectable residue levels, nor does it have any metabolites, without sufficient documentation. The purpose of GLPS is to create thorough documentation. The conduct of the study and the level of quality control measures will then be evident through the auditing process.

In conclusion, the quality of the science, or good science if you perfer, can be considered a myth since

individual value systems are part of developing a
quality program. The GLPS require thorough
documentation of a study to justify the quality and
integrity of the work. The chemist's objective should
be to obtain valid measurements and then be able to
support the findings and conclusions of the study
through documentation. Quality in the form of good
science and quality in the form of compliance to GLPS
and documentation are not the same and therefore a claim
to quality can not be made based solely on the existance
of a GLP program. Supportive documentation is however
the key to compliance and therefore the key to quality.

RECEIVED February 19, 1988

Chapter 8

Standard Operating Procedures

One Element of a Program for Compliance with Good Laboratory Practice Regulations

Alice E. Parks

Agricultural Products Department, Experimental Station,
E. I. du Pont de Nemours and Company, Wilmington, DE 19898

Standard operating procedures (SOPs) are documents
specifying procedures that must be followed to assure
the quality and integrity of study data. SOPs are
intended to reduce the introduction of errors and
variables in a study by assuring that appropriate
procedures are used consistently. They have a
historical purpose after completion of a study as
documentation of the procedures that were used. SOPs
are one element of a compliance program required by
the Good Laboratory Practice (GLP) regulations for
studies that are submitted to support the registration
of pesticides regulated by the Environmental
Protection Agency (EPA). GLPs also apply to studies
submitted under the Toxic Substances Control Act.
This paper provides an overview of SOPs including
regulatory requirements, and guidelines for
establishing and maintaining a system of SOPs.

The Good Laboratory Practice regulations are basic principles that
have been developed to assure the quality and integrity of data
generated from studies used for hazard assessment. These principles
address the general processes for conducting studies, documenting
procedures and results, and retaining records. They do not address
scientific considerations.
    In 1979, the first GLP regulations became effective for
nonclinical studies submitted to the Food and Drug Administration.
Since then, an increasing number of study types required by
regulatory agencies for hazard assessment have been required to
comply with GLP regulations. The EPA, under the Federal
Insecticide, Fungicide and Rodenticide Act, will be proposing
generic GLP regulations to address additional types of studies
submitted for registration/reregistration or experimental use
permits of pesticide products. The studies included will be
environmental fate as well as certain product chemistry and
ecological effects studies, among others. These generic GLPs will

0097-6156/88/0369-0047$06.00/0
© 1988 American Chemical Society

apply the same principles established by the first GLP regulations. With this expansion comes challenges for novel applications of GLP principles to suit the different types of studies. Some of the greatest challenges will be presented by field studies because of the large number of studies conducted in numerous distant locations. Individuals with skills in Quality Assurance (QA) will be working closely with scientists who are familiar with these studies to develop compliance programs.

Standard operating procedures are one aspect of a complete GLP compliance program. They are relatively straightforward to apply to these additional study types. The following review of the GLP requirements for SOPs and recommendations for developing an SOP system are presented to help individuals who are learning the GLP principles and/or developing a GLP compliance program.

Definition and Purpose of SOPs

Standard operating procedures are written documents specifying the procedures that must be followed to assure the quality and integrity of study data. One of the purposes of SOPs is to reduce the introduction of errors and variables in a study by assuring that appropriate procedures are used consistently by all personnel. The other purpose, which is a historical perspective, is to outline how a study was conducted.

The distinction between SOPs and protocols is often unclear to individuals who are becoming familiar with the GLP regulations. A protocol outlines the objectives and methods for a study; it indicates what will be performed during a study. In contrast, SOPs are more specific and outline how the portions of the study will be conducted. For example, a protocol for a cow residue study outlines the tissue samples to be taken at necropsy, whereas an SOP outlines the procedure for collecting the samples.

Value of SOPs

The use of SOPs results in additional benefits to an organization beyond GLP compliance. Some of the ways in which they are beneficial are outlined below.

- SOPs outline the critical aspects of a procedure and help to assure that these aspects are appropriately emphasized during the conduct of the procedure.
- SOPs help to assure consistency among individuals who are performing a procedure. As outlined previously, they help to prevent the introduction of errors and variables. When they are in place, individuals do not have to rely on memory or word-of-mouth communication of procedures.
- SOPs can help assure that appropriate documentation and data collection occur by outlining the records to be generated during the performance of a procedure.
- SOPs can be used in training individuals to clearly communicate the specific method for performing a procedure. This helps to prevent misunderstandings.
- SOPs help to assure that personnel perform work according to the most up-to-date standards or methods that are outlined.

In this way, they prevent confusion in identifying the most current method to be used.
- SOPs improve planning and organization. Preparing an SOP requires an individual to think through the process to be described. In doing this, potential problems can be identified and eliminated.
- SOPs assist in the effort to standardize, and this improves efficiency. A procedure that is written as an SOP eliminates the need to redevelop the procedure each time it is performed.
This aspect is particularly helpful to organizations that are starting a GLP compliance program. SOPs that address procedures such as preparing and amending protocols and final reports, help to ensure that standard practices are used in the organization.
- When contracting work, sponsors can review SOPs from the contract facility to understand the specific procedures that will be performed during a study.
The end result of these benefits includes improving the accuracy and integrity of data generated, improving communication, and improving efficiency. In a worst-case situation, SOPs can help in preventing the repetition of part or all of a study and can aid in preventing significant errors that might otherwise remain undetected.

An example of a situation demonstrating how SOPs can add value to an organization by preventing wasted effort is presented below. After our organization introduced SOPs for field residue testing, the individual responsible for a previous application of test mixture was checking the procedures he had used for calculating the amount of material applied with the method outlined in the SOP. After checking his calculations, the individual discovered he had miscalculated and that the crop already harvested from this plot would not meet the objectives of the study. The samples had to be discarded.

Guidelines for Preparing SOPs

An appropriate SOP should exist prior to performing each procedure during a study. Each facility or group should establish a method or system for organizing SOPs, determine the general content and assign personnel to write them. This can make the process of preparing, using, updating and retaining SOPs most efficient.

Organization of the SOP System. In planning the preparation of SOPs for the many procedures that will be performed in the facility, a method for managing the SOPs should be considered first. The organization of these documents should allow for maximum ease in the following areas:
- Use - Most importantly, the SOP system should be organized to ensure ease and efficiency in use.
- Preparation - SOPs should be prepared and inserted into the system (or eliminated) at any time without rearranging the whole system.

- Referencing - SOPs should be indexed or numbered for easy reference.
- Revision - Revisions should occur in a timely manner and with minimal impact on the system.
- Reconstruction - On a historical basis, personnel must be able to identify the SOPs that were in effect for any study.

Independent SOP subdivisions or units should be written rather than a "text" of SOPs. A separate SOP for each function, or for each model or type of equipment should be prepared.

An indexing or numbering system allows for ease in referencing and locating the SOP for current use, and it assists in pinpointing specific SOPs that were in effect for past studies. A unique number or alphanumeric code for each SOP can meet these needs.

The indexing system should be structured in a manner that is consistent with the organization of the facility or with the type of work conducted. For example, the first level of indexing can be based on functional groups, such as the QA Unit. In addition to SOPs for each functional group, a general category of SOPs can be developed to address procedures common to all groups. Within each functional group, the next level of indexing can distinguish between procedural and equipment SOPs. Each SOP can be identified further with a unique number. A final level for the indexing system can designate the revision number for each SOP.

Content of SOPs. In preparing SOPs, these key points should be followed:

- Use a clear and descriptive title for each SOP.
- Outline the critical aspects of performing the procedure to ensure that it will be conducted correctly and to ensure that the data generated is of high quality.
- Provide sufficient detail without being unnecessarily restrictive. The SOP must meet the need of an individual user while being general enough such that it is appropriate for more than one user. Flexibility should be written into an SOP whenever appropriate; however, if an SOP is too general, it may be useless in meeting its intended purpose.
- Organize the SOP by ordering the sequence of events involved in performing the procedure. Present the text in a straightforward and easy-to-follow manner. After drafting the SOP, use it in performing the procedure or operating the equipment to ensure that it is clear and has sufficient detail to be followed by trained personnel.
- Published literature (e.g., textbooks and manuals) may be referenced in an SOP or may be used as a supplement to an SOP. However, published literature alone does not completely address the specific needs of a group or facility. Publications usually contain more information than is appropriate and are not clear enough in specifying which procedure to use.
- If forms are used in collecting data when performing the procedure, they may be included. Alternatively, the title of the form can be referenced and current forms can be located in a file. This latter option facilitates updating forms without revising SOPs.

- Indicate the effective date of the SOP.
- Indicate the total number of pages so the users can be certain that they are performing the complete procedure.
- Have the preparer(s) and management sign each SOP.

In addition, GLP regulations require SOPs for equipment to address the following specific items:

- Methods, materials, and schedules to be used in the routine inspection, cleaning, maintenance, testing, calibration, and/or standardization of equipment.
- Remedial action to be taken in the event of failure or malfunction of equipment.
- The person responsible for the performance of each operation. The person's position or title should be used rather than a specific name to avoid unnecessary SOP revision when a person changes responsibilities.

Assignment of Personnel to Prepare SOPs. One approach in assigning responsibility for preparation is to involve at least one person from each work area of the organization to write and revise SOPs for the area. Consistency is reached by having a person coordinate the SOP system and the individuals from the different areas. A guidance document for preparing SOPs helps to establish uniformity in the system and helps to ensure each SOP contains appropriate information. It is particularly useful to organizations that are developing their compliance program.

Along with helping to promote cooperation among personnel, this approach ensures that these documents are technically sound, and that individuals who conduct the work/studies are responsible for the accuracy of the SOPs and for following the SOPs.

Responsibilities Associated with SOPs

The GLP regulations outline responsibilities of different individuals or groups pertaining to SOPs that include:

- Management
  Management must be satisfied that the procedures outlined in the SOPs assure the quality and integrity of data that are generated during studies, and should sign the SOPs to document this. In addition, management must authorize in writing any significant changes in established standard operating procedures.
- Study Director
  The study director is responsible for ensuring that SOPs are followed and that deviations are documented in the raw data.
- User
  Individuals performing the studies are responsible for following the SOPs and documenting deviations from SOPs.
- QA Unit
  Through auditing, the QA Unit must determine if SOPs are used and that deviations from SOPs were properly authorized and documented.

Revisions to SOPs

SOPs are subject to continuous revision reflecting influences from many sources such as changing technology, efficiency improvement of methods, etc. An SOP should be revised as soon as the procedure is identified as permanently changed. As outlined previously, management must authorize in writing all significant changes to issued SOPs. To ensure that SOPs are current and accurate, they should be reviewed on a periodic basis (e.g., annually) by personnel who are assigned to prepare them. A method should be developed to ensure users are aware of current versions.

The study director must authorize nonpermanent deviations from SOPs that are specific for a particular study. These do not require an SOP revision. The deviations must be documented in the study records.

Location of SOPs

SOPs that are pertinent to the work conducted in each area must be immediately available so that individuals have direct access to them and can ensure that they are accurately performing the procedures. If SOPs are not readily available, people are unlikely to retrieve them due to factors such as time pressures or inconvenience. Published literature that is used to supplement SOPs, such as an instrument manual, also must be readily accessible.

In addition, the QA Unit should have a copy of all current SOPs to enable QA personnel to refer to them when auditing.

Retention of SOPs

All SOPs, including current and obsolete versions, must be retained. This enables future reconstruction of the procedures used to perform any study, independent of personnel who were involved in the study. Documenting the specific SOPs used in the study records clearly identifies the appropriate SOPs.

One group in the organization should be assigned the responsibility for maintaining this file of SOPs. An indexing system and dates indicating when the SOP was effective are necessary to ensure proper identification of SOPs in minimal time.

EPA Inspection of SOPs

During a GLP inspection of a facility by the EPA, the inspectors usually examine SOPs and may check the items outlined below to determine the compliance status of the system. Commonly, inspectors request copies of SOPs to include them in the inspection report in support of their observations.

- If a data audit is performed, the inspectors may request to see the SOPs that were in effect when the study was conducted.
- To determine if current SOPs exist for appropriate functions, inspectors may request a list of the SOPs for the organization. Additionally, they may review the content of specific SOPs.

● While touring the facility, the inspectors will probably ask to see SOPs that are immediately available for the procedures performed in each area.

● To evaluate the user's knowledge of SOPs, the inspector may direct specific questions to the user. For example, the inspector may ask an individual where a specific SOP is located while the inspector is in an area where that procedure is performed, or the inspector may ask the individual how he/she becomes familiar with new or revised SOPs.

Examples of Topics to be Addressed in SOPs

Some examples of topics to be addressed in SOPs for studies such as environmental fate and residue are outlined below.

● Chain-of-custody procedures for test substances or mixtures - including procedures for receipt, storage and distribution
● Preparation of test substances or mixtures for application
● Application of test substances or mixtures - including procedures for preventing cross contamination
● Equipment use, maintenance and calibration
● Collecting samples - including procedures for taking representative samples, preventing cross contamination among samples, and identification of samples
● Procedures for chemical analysis - including procedures for analysis of test substances, application mixtures and samples
● Chain-of-custody procedures for samples - including procedures for storing, packing and shipping
● Archiving test substances, study records and samples
● QA procedures - including auditing procedures and maintaining the master schedule
● Computer use - including procedures for data generation, validation of software, security, etc.

A relatively complex area to address is procedural SOPs for chemical analyses in metabolism studies. One approach is to address the major operations that are common to the studies, for example characterization of metabolites in soil. The SOP can describe the general process, options available in the process and requirements for acceptance or rejection of data. Study-specific procedures that complement the SOPs can be outlined in detail and retained as part of the study records. These study-specific procedures can be prepared in the form of a work sheet and used for entering original documentation, such as the person who performed the procedure and the date it was performed.

Summary

Expanding the GLP regulations to include additional studies that are submitted to regulatory agencies requires applying the basic GLP principles to new areas. The purpose of these principles is to assure and document high quality data for hazard assessment. SOPs contribute as an element of GLP compliance by helping to assure that the appropriate procedures are consistently used in performing studies.

In implementing SOPs as part of a compliance program, each organization should develop a system for managing the SOPs that will be effective for its operations. When preparing specific SOPs, the organization should focus on the purpose of SOPs, the regulatory requirements, and the recommendations presented above. This will maximize the benefits that can be realized through the use of SOPs.

RECEIVED January 29, 1988

Chapter 9

The Protocol and Its Impact on Research Activities

Robert J. Daun

Hazleton Laboratories America, Madison, WI 53704

Existing Good Laboratory Practice regulations (GLPs) mandate that each study have a protocol that clearly defines the key elements of the study. Although the GLPs define the key elements of a valid protocol, the document must be designed to provide a proper balance between an exact definition of what will be done and still retain a degree of flexibility. Although a protocol is required by GLPs, it should never be regarded as a document prepared solely to meet administrative requirements. Rather, it should be considered a necessity to ensure an efficiently conducted scientific study of high quality. The purpose of this article is to discuss the impact of a well-designed protocol on the performance of the study. Examples are given as to how the protocol is used, both before study initiation and during the actual conduct of the study.

Existing Good Laboratory Practice (GLP) regulations dictate that "Each study shall have an approved written protocol that clearly indicates the objectives and all methods for the conduct of the study."(1) The term "protocol" has become a key word in the vocabulary of scientists involved in health-effects studies since the late 1970's. To many, the word protocol itself has taken on a new and highly narrowed meaning. An examination of the dictionary definition of the term protocol provides a point of interest not only in the historical derivation of the term but also some unique insights in how the document should function in the current context of a regulated study. A part of Webster's(2) definition of the term protocol is ". . . first sheet of papyrus roll bearing the authentication and date of manufacture of the papyrus. Papyrus roll, sheets of papyrus glued together, literally, that which is glued together." Within this classical definition can be found some key items relevant to the use of the term protocol. These two items are "authentication" and "glued together." The study

0097–6156/88/0369–0055$06.00/0
© 1988 American Chemical Society

protocol, as mandated by existing and proposed GLPs, serves as an
authenticating document bearing the signatures of key parties to be
involved in the proposed study; and in the current vernacular,
serves to "glue together," in an orderly fashion, the specific
operating requirements of the study.

The format of a study protocol is left to the entity
developing the protocol; however, the basic requirements are
clearly set forth in the GLP regulations. The forthcoming broad
application of GLPs to chemistry-based and field studies means
that, for the first time, many groups will find the need to develop
study protocols that will conform to the requirements of GLPs.
There needs to be a recognition by the study scientists that
development of this document is much more than an administrative
exercise. This document, when properly constructed, provides the
key working tool for the successful completion of the study.

The Protocol--What Does it Provide?

The study protocol must, first and foremost, contain a clearly
stated objective of the research activity to follow. It has not
been unusual for studies to be conducted without all of the study
team understanding the scope of the investigation. For example, a
study conducted to determine the identity and relative quantity of
a pesticide and its metabolites in the edible portions of
food-producing animals should be restricted to the activities
necessary to provide this information. Without a clearly stated
objective, this type of study could instead be manipulated into an
attempt to determine toxicological responses or pathological
effects. The data obtained could be of questionable value because
the protocol design would not contain the necessary elements to
provide reliable data. This is not to say that combined studies
are of no value, but if a study is multidisciplinary in nature, the
study design should contain input from staff qualified in the
disciplines involved.

A protocol provides a focus of responsibilities for all
members of the study team. The combination of a clearly stated
objective, along with a clear definition of the methods and
materials, will avoid confusion during the study when often there
is not time for extensive consultation among the study team.

Most studies intended for submission to a regulatory agency
are developed to correspond to the requirements of some type of
guidance document, such as the Federal Insecticide, Fungicide, and
Rodenticide Act (FIFRA) guidelines for pesticide registration
studies. By its very nature, a guideline needs to be flexible
enough to allow for variation in a specific study. Because each
study must be tailored to the specific compound under investigation,
it is wise to solicit regulatory agency review of the study protocol
to verify that the study will provide the required information. If
reviewed and approved, the protocol can provide a means to avoid
costly and time-consuming study repeats.

Whether a study is performed within the sponsoring facility
or by a contract laboratory, it is necessary that there be a review
by all parties involved. In addition to review by the study
director and, in the case of a contracted study, the Sponsor's

study monitor, it is wise to have the protocol reviewed by the
Quality Assurance Unit to verify compliance with GLP protocol
requirements. In the case of a study that involves radioactive
isotopes, the protocol should be reviewed by the performing
laboratory's Radiation Safety Officer to ensure that all
requirements necessary for worker safety and prevention of
inadvertent contamination are specified. The signed final protocol
provides written documentation that all parties are in agreement as
to the conduct of the study.
    The well-constructed study protocol provides a ready reference
for study-specific information. The final protocol should be sent
to all of the key members of the study team and a copy should be
present at all sites where any activity associated with the study
is being performed. Even though the protocol may have been
reviewed by the study team, it is not unusual for questions to
arise during the course of a study that can be readily resolved by
another perfunctory review of the protocol wording.
    The study protocol provides a vehicle for estimating the time
and costs of a study and makes the study team aware of any unusual
scheduling or non-typical equipment required to carry out the
study. Although the use of the protocol for costing purposes is of
extreme importance to a contract laboratory, the same information
is of importance for noncontracted studies. It allows the
performing facility to ensure that all of the required resources,
in terms of trained staff, facilities, and equipment, will be
available and in place at study initiation.
    A protocol provides a mechanism for review of data and
reports during, and at the conclusion of, the study. A regular
review of the actual study conduct, in reference to the planned
conduct as detailed in the protocol, provides a degree of
confidence that the results obtained will be consistent with the
study objectives. That is not to say that minor deviations can
not, or should not, occur throughout the course of the study. The
GLP regulations, however, require that these deviations be noted in
the final report of the study. The study protocol provides the
"master" reference for compilation of these deviations.
    More than anything else, the protocol is a working document.
It is the single most important reference for addressing common
questions that arise during the study. As a working document, it
is necessary that multiple copies be available for use by members
of the study team. Once the document is finalized with all the
appropriate signatures, it is important that it not be relegated to
a file where it is not readily available when needed.

The Major Impacts of a Well-Designed Protocol on Research Activities

One of the major impacts of a well-designed protocol is in the area
of prestudy planning. It is a common occurrence in the planning of
a study, that a number of changes to the initial or draft protocol
will be dictated. In this fashion, the planning process comes full
circle; the draft protocol provides a planning document, and the
planning process itself will result in a final protocol requiring
few amendments or significant deviations.

A well-designed protocol also produces efficiency of effort. Studies are normally budgeted to be run one time without extensive study restarts or repeats. In addition to the wasted personnel efforts from repeated studies, submission deadlines are often severely compromised, either delaying the introduction of new products or risking the continued registration of existing products. A well-designed study described in an accurate and thorough protocol is often the key to an efficient and productive research effort.

The protocol is the key item in avoidance of confusion and misunderstanding among the various parties involved in a study. Although this is of prime importance in cases where two entities (e.g., the Sponsor and a contract laboratory) are involved, the same potential for problems exists within an internal study. In some cases the potential may be greater with an internal study because potential communication difficulties are more readily anticipated and addressed then when dealing with a contract laboratory geographically far removed. Some recommended prestudy practices that involve the use of the study protocol will be discussed later.

A final impact of the protocol is pertinence. It allows for administrative review to ensure that the study, as designed and described, pertains to the needs at hand; and that the result will not only meet the stated study objective but also fulfill regulatory agency requirements.

Some Factors to Consider in the Design of the Protocol

A well-designed protocol presupposes a well-designed study. It is possible to design an accurate and thorough protocol, complete with objective, that does not produce the required information. The major problem in this case could be that the parameters of the study would not completely encompass the needs outlined in agency guidelines. For example, application rates for a field residue study may not be properly selected, or dosing levels or sacrifice intervals may not be proper for a study that uses domestic or laboratory animals. As mentioned previously, if questions exist, the protocol should be submitted to the appropriate regulatory agency office for review.

Ambiguity--there should be no statements that might require interpretation by the scientific staff during conduct of the study. This applies particularly to those factors where numbers are critical. Examples are dose rates or application levels, sacrifice or harvest intervals, and replication requirements. The protocol, however, must also retain a degree of flexibility in those areas where exact definition is not needed or cannot be determined prospectively. Specifications should not be so detailed that there is no allowance for equivalent substitution. Usually, it is not necessary to specify brand names; however, there may be times when experience dictates that a specific brand or manufacturer are required to perform a given function. In those cases, of course, specificity is not only desirable but mandatory.

In designing or developing the study protocol, it is wise to minimize assumptions. The fact that similar studies have been done previously should not be used as a reason for not clearly defining the operations to be done. It may be necessary to replace some members of the study team shortly before initiation, and their ability to review and assimilate the necessary study details can be critical to success.

In order to make the protocol an effective working document, it is desirable to keep statements as concise and direct as possible. Avoid excessive verbiage. Excessive use of descriptive text only makes it difficult to quickly find the answer to questions. The appropriate use of well-defined and clearly spaced "headers" allows staff members to locate the pertinent section of a protocol quickly without the need to read many pages. Many headings can be completely and accurately defined by a single word or short phrase without the need for a complete sentence.

Another technique that can be employed to increase the ease of use of the protocol is to establish a format and then keep the format as uniform as possible from one study to the next. In this way, members of the study team will develop familiarity with it and will be able to easily find the appropriate page or section of the protocol for resolution of questions. An added benefit to maintaining consistent formats is found with the people responsible for generating the documents. The time required to do this necessary part of the operation can be significantly reduced with less chance for errors.

Some Recommended Practices in the Use of the Study Protocol

One of the most useful practices to establish in any facility is the pre-initiation conference. The pre-initiation conference is a face-to-face meeting of all study team participants, essentially to do a line-by-line review of the protocol. It is at this time that questions or clarifications should be brought forward. During the pre-initiation conference the study team participants may discover that material in the protocol is not adequately defined. The conference should be scheduled far enough ahead of study initiation to allow for preparation and dissemination of a revised protocol before study initiation. In cases where the study is to be preformed by a contract laboratory, a visit by the Sponsor's study monitor at the time of the pre-initiation conference is highly desirable. If this is not possible, the conference must be scheduled early enough to allow Sponsor input into protocol modification. The person chairing this conference should make a point of notifying the study team of any operations that are new or that differ from the usual study procedure.

Another practice that has been found to be extremely useful is the clear designation of the final protocol. Many times a draft protocol will go through several minor revisions before it is finalized. This can then result in study team members having a version of the protocol that does not incorporate all of the final changes. One way to avoid this potential problem is to print the final protocol on a specific-colored paper thereby distinguishing it from earlier versions. Any further changes are then made by

amendments that, when signed, are also copied to the specific-
colored paper and become an integral part of the working document.
As mentioned earlier, the protocol should be readily
available to all personnel involved in the study for prior review
and for use during the actual conduct of the study.

General Comments
If techniques are employed that are new or non-routine to some
members of the study team, these need to be expanded on in the
protocol. As stated earlier, minimize assumptions.
Do not include items that are routine if they are clearly
defined by internal Standard Operating Procedures (SOPs).
Reference should, however, be made in the protocol that these items
are SOP driven. Examples might be subjects such as safety
procedures, wearing apparel, and handling use and disposal of
radioisotopes.
Do not include brand or trade names unless necessary to the
successful conduct of the study.
Avoid a simple regurgitation of agency guidelines. The
guidelines are there for assistance in designing a study to meet
certain scientific requirements. The protocol and subsequent study
should be designed using scientific judgment appropriate to the
specific test material to be investigated.
Ensure that the protocol meets the requirements for
compliance with GLPs, and scientifically addresses guideline
requirements. It is also important to be aware that in some cases
more recently published reporting addenda and standard evaluation
procedures contain information critical to proper study design, and
therefore, protocol developement. If there is a question, check
with the regulatory agency.
It can be seen that the protocol is much more than a document
necessary to meet an administrative requirement. Proper
development and deployment are the keys to providing a valid
scientific investigation. Outside of a well-designed study, there
is no single document or factor that can be as crucial to the
success of a study as the protocol.

Literature Cited

1. 40 CFR 160.120.
2. Webster's Third New International Dictionary, Babcock Gove, P.
    Ed., Merriam-Webster, Inc.; Springfield, Massachusetts 1986;
    p. 1824.

RECEIVED January 29, 1988

Chapter 10

Raw Data Definition and Documentation

Edward J. Panek

Agricultural Research Center, Chemicals Division, BASF Corporation, Research Triangle Park, NC 27709

The current EPA Regulations (40 CFR 160) require that "all raw data, documentation records, protocols, specimens, and final reports generated as a result of a study shall be retained". The types and amounts of raw data generated in agrochemicals research are illustrated using a small planned field residue program as an example. This example is also used to illustrate documentation of this data using hierarchical paper files and relational electronic data base files. The archival needs for storage of this data are also given.

This paper is concerned with the definition and documentation of primary raw data, or in other words, raw data directly associated with a study. Items such as standard operating procedures, methods, personnel qualifications and training records can be considered secondary raw data. Thus, even though these items are also archived, they are not considered explicitly here. But since they are just other examples of paper and/or electronic records, the same archival methods can be used.

Most definitions of raw data concentrate on paper and/or electronic records. A good, concise definition is found in the FIFRA Good Laboratory Practice document (1): " 'Raw data' means any worksheets, records, memoranda, notes, or exact copies thereof, that are the result of original observations and activities of a study and are necessary for the reconstruction and evaluation of the report of that study. In the event transcripts of raw data have been prepared (e.g., tapes which have been transcribed verbatim, dated, and verified accurate by signature), the exact copy or exact transcript may be substituted for the original source as raw data. 'Raw data' may include photographs, microfilm or microfiche copies, computer printouts, magnetic media, including dictated observations, and recorded data from automated instruments."

Another major type of raw data is samples. These are mainly retained aliquots of test chemicals and the biological samples generated in field residue trials, metabolism studies, and environmental fate studies. The chemical samples are to be archived

0097–6156/88/0369–0061$06.00/0
© 1988 American Chemical Society

for the life of the registration or for as long as the quality of
the preparation affords evaluation. The biological samples are
normally retained only until the analysis results are verified
(audited) and/or until the known storage stability is reached.
Nevertheless, the archival challenge represented by the large volume
of these biological samples is only slightly diminished by their
shorter storage time.

The types of archival storage facilities needed are functions of
general archival needs and the types of items to be archived.
General archival needs are controlled access, safe and appropriate
storage, and retrievability. The types of items to be archived
include paper, field samples, test chemical samples, and electronic
records. In general the following types of archive facilities are
needed: sample freezers and cold rooms, test chemical freezers,
file cabinets and/or boxes for paper, and magnetic media for
electronic records storage.

Raw Data from a Planned Field Residue Program

The types and amounts of raw data generated will be illustrated by
following a planned field crop residue program from its definition
(protocol) stage to the final report. The archival/documentation
needs will then be similarly illustrated.

Protocol. The first step is to define the program (i.e., write the
protocol). Key technical elements to define are: 1) the crop or
other use, 2) the use rate(s), timings of applications and
application techniques, 3) the test chemical and its formulation,
4) raw agricultural commodities to harvest and the schedule, and
5) test locations. Each of these elements generates information,
i.e., raw data. The choices of general test locations and the
harvest commodities for each crop are defined by USDA and US EPA
information (2-3).

Test Chemical. Once the test chemical and its formulation have been
defined, the needed amount is prepared and packaged, the batch is
analyzed, and portions are shipped to the test locations. Retained
samples are correctly stored. If the storage stability of this
formulation has not been determined, some of the retained samples
are used to determine its storage stability under typical storage
conditions. In this and subsequent data gathering steps the
investigators collecting the data need to be identified.

Field Locations. For each field test location a variety of information is collected and recorded in addition to the samples which are
collected. Seven general categories of information can be defined.
Three simple ones are: 1) test design or plot plan, 2) location
and 3) field use history for several years. Field soil characterization 4) includes screen analysis (soil type), pH measurement,
and organic matter content. Weather information 5) includes daily
temperatures and rainfall and/or irrigations during the test.
Application related data 6) consists of dates, application modes,
weather conditions at application, calculations and calibrations.
Harvest information 7) includes crop name, part, amount, date, and
collector.

Field Samples. Each field sample is packed in an appropriate
container and labeled. In addition to the harvest information
listed above, information on the history of the sample (storage
conditions and intervals) near the field location and on shipment
to the laboratory is generated.

Analysis Information. As with each field test location, for each
sample or set of samples a variety of information is also generated
in the laboratory. This information can be grouped into four
general categories. Sample handling records 1) include receipt
condition, processing and sub-sampling, storage conditions and
sample access information. Analysis procedure records 2) include
the sample sizes, aliquoting, dilutions, etc. for the method of
analysis used. These methods usually contain extraction, clean-up,
and derivatization steps. The analysis method is applied to treated
samples, control samples and method recovery samples (spiked control
samples). These records may be in the form of bench sheets or
laboratory notebooks. Chromatographic information 3) includes
the actual analyses of the samples mentioned above, as well as the
information on injection standards (standardization or calibration)
and the instrument log books. Calculation information 4) shows how
the analysis procedure and standardization data are used with the
chromatographic data to determine test chemical (and degradation
product) concentrations in the harvested commodities.

Final Report. The final report summarizes all of the above
information.

Amounts of Raw Data. The types of paper and/or electronic records
generated in this example are shown in Table I.

Table I. Types of Records Generated

Category	Number
Test Chemical	3
Field Location-Each	7
Field Samples-Each	2
Analysis-Each Sample	6

These consist of field data sheets, sample storage records, bench
sheets or laboratory notebooks, chromatograms, and shipping papers.
Some of the laboratory records, in particular, can be electronic
rather than paper.

The amounts of each type of record generated in a planned field
residue program depend on the number of test locations and the
number of raw agricultural commodities harvested. In a small (10
locations) simple (2 commodities harvested) program, 40 samples
(20 treated, 20 controls) are harvested and approximately 170
direct records are generated. Most of these records consist of
multiple pages so that approximately 400 pages (or equivalent) of
records are created for this program. The following equation
dramatizes this point.

$$\frac{\text{Weight Records}}{\text{Weight Sample}} = 10$$

(It is actually more true if number is substituted for weight in the above equation.)

Documentation/Archival Needs

The previous sections indicated the types of items and their amounts generated in each of the major steps in a planned field residue program. In this section archival needs related to these items are given. These are grouped by archival requirement.

Archive Management. An individual must be responsible for the archives. This person controls access to the archives, checks items in and out of the archives, and maintains these use (access) records.

Controlled Access. Access to the physical archives (e.g., file cabinets and freezers) is controlled by locks and the archive management. Access to electronic files is controlled by secret user identification (ID) numbers. Well designed electronic data storage software records or stores the ID number of any user that enters or changes data and when that entry or change occurred.

Storage Conditions. Storage conditions are designed to minimize deterioration of the archive contents. Since the contents differ greatly, so must the optimum storage conditions. Chemical samples are typically archived in freezers. Biological samples are stored in cold rooms or freezers. Paper or microfilm records are stored in cool areas where the chances for fire and light caused damage are minimized. Electronic media are stored under the above conditions in the absense of strong electrical or magnetic fields. Magnetic tapes need to be backed-up (remade) periodically.

Retrieval Methods. The magnitude of records created in our small example program clearly establishes that the heart of any documenta-tion or archival system is the systematic retrieval of specific items. This is one of the strengths of electronic data systems. Thus the key information on our paper records is also contained in electronic files.
    These cross-referenced numbers are the key to the electronic relational data bases. Key field data and sample storage data are entered into location and sample number files in QUIZ Software (4). Laboratory analysis information is contained in files generated using Perkin-Elmer Laboratory Information Management System (LIMS) and Chromatographic Laboratory Analysis System (CLAS) software. Both of these systems have magnetic tape back-ups for the hard disks.
    The hierarchical paper data file is organized in the same way the planned field residue trial example was developed. The protocol and final report are followed by test chemical information, test location information, application and harvest information, shipping and storage data, and analysis data. The analysis data is grouped by bench sheet to assist manual searches.

Conclusions

The types and amounts of raw data generated in agrochemicals research were illustrated by using a small planned field residue trial as an example. The large amount of raw data generated in this small example indicates how throughly both the study and the archival storage must be planned. The diversity of materials to be archived also contributes to the complexity of the archival needs.

Acknowledgments

To BASF Corporation Chemicals Division for permission to publish this paper and support during its writing at the North Carolina beaches.

Literature Cited

1. Code of Federal Regulations (40 CFR), Part 160, Good Laboratory Practice Standards.
2. Agricultural Statistics, U.S. Department of Agriculture, U.S. Government Print Office: Washington, DC, annual publication.
3. Schmitt, R.D. Pesticide Assessment Guidelines Subdivision O: Residue Chemistry, U.S. Environmental Protection Agency: Washington, DC, 1982.
4. A product of Quasar Systems Ltd., Ottawa, Ontario.

RECEIVED February 2, 1988

Chapter 11

Computer Systems Validation

How To Get Started

Ronald C. Branning

Boehringer Ingelheim Pharmaceuticals, 90 East Ridge,
P.O. Box 368, Ridgefield, CT 06877

The proliferation of computers in the production of
pharmaceuticals resulted in the U.S. Food and Drug
Administration (FDA) publishing the "Guide to
Inspection of Computerized Systems in Drug Processing"
in 1983. FDA Inspectors have been using this
guideline for the past three years to cite firms for
their failure to validate their computer systems.
Other U.S. regulatory agencies are now asking for
validation of computer systems in chemical, R&D, and
clinical inspections. This presentation will briefly
review the U.S. regulatory posture and industry
response concerning computer systems validation and
will review in detail a practical step-by-step
approach to identifying, classifying, validating, and
documenting computer systems.

Computers are involved in virtually every facet of modern life.
Their application to the production of pharmaceuticals prompted the
Food and Drug Administration to publish the "Guide to Inspection of
Computerized Systems in Drug Processing", The Blue Book, in 1983.
The Pharmaceutical Manufacturers Association's Computer Systems
Validation Committee answered the FDA's document with "Validation
Concepts for Computer Systems Used in the Manufacture of Drug
Products" in 1985. Several authoritative papers have also addressed
this subject; they are listed in the references at the end of this
article. The purpose of this paper is to outline a practical
approach to implementing the recommendations from these sources.
Although this approach was developed for use in GMP regulated
pharmaceutical firms, it will work regardless of the compliance
guidelines being used. The formation of the management team, the
identification of "validatable" computer systems, the definition of
documentation requirements, and the development of validation
protocols are the key points covered in this article.

Computer systems validation is not a new, magic formula. The
techniques are the same ones used in any structured approach to
project management. Unfortunately, when the word "computer" is used
in conjunction with a topic, it suddenly becomes shrouded in a veil
of mystery. You can lift this veil over computer systems validation
by following these steps, one at a time.

Initiation of Computer Systems Validation

The first step is for someone in management to recognize the need
for computer systems validation and to gather the other management
expertise necessary to address the issue. This person is usually
someone in the Quality Assurance Unit (QAU) or the computer
operations, Management Information Systems - MIS, group. While
these two departments need to be involved, computer systems users
such as laboratory leaders, study directors, and other affected
department managers need to be included in the development of a
computer systems validation plan.

Steering Committee. Although committees have a reputation for being
inefficient, a properly structured committee approach may be the
most effective and efficient way to approach the relatively complex
process of computer systems validation. The complexity stems from
the necessity for a multidisciplinary approach to validation not
just from the fact that computers are involved. The chairmanship of
the core group should be from either MIS or QAU since MIS is the
most involved in the technical aspects of validation and QAU is the
main regulatory contact concerning validation. The steering
committee in this scenario represents the policy-making board. The
computer systems validation policy, resource allocation, and final
validation approvals are the responsibilities of this group. The
committee should be formed at the director level so that direction
can be determined, necessary resources allocated and final decisions
made. Unfortunately, this level is usually too far removed from the
actual systems being validated, therefore an operating committee
should be formed.

Operating Committee. A working committee at the manager or level
within each definable and logical business group is needed to
develop the SOP(s), write and review the protocols, keep the
validation projects on track, make the day-to-day decisions
regarding individual system validation problems, and raise the
unresolved policy issues to the steering committee. Once again, the
chairmanship of the operating committee should be from either MIS or
QAU, mirroring the steering committee. The membership of the
operating committee should be kept to a minimum with adjunct
membership of user representatives as needed. In small operations,
both of these committees' functions can be handled by one group or
even one person wearing several hats.

Responsible Users. The actual work of following the SOP's,
developing the protocols, executing the test plan and summarizing
the data is done by "responsible users". Responsible users are the
manager/supervisor level people who have control of the computer
operations. In this scenario the responsible user is the project
manager reporting to the operating committee; computer operations
(MIS) is the technical support to the responsible user in the
validation effort with back up from the operating committee.

Documentation

The primary tasks of the operating committee are to develop an
operating procedure (SOP) and a validation protocol outline. The
SOP should be the "what to do"; the protocol the "how to do it"
including a listing of required documentation.

SOP

Writing an SOP is usually a task for one person working with a group
of advisors. In this case, one member of the operating committee
should be assigned the task with support from the other committee
members.

Objective and Scope. The objective and scope of the SOP need to be
carefully thought through and described. A limited objective could
be to validate only those computer systems directly related to the
production of pharmaceuticals; the broadest one would be to validate
all computer systems regardless of their application. Usually it is
somewhere between the two. The scope will be determined by the
company philosophy, organizational structure and the number of
divisions, plants or departments involved. The scope should be
limited to as small a unit as possible for the initial validation
effort in order to achieve at least one successful computer system
validation quickly.

Definitions. Each business group will have a set of working
words and definitions to describe computer systems and their
operations and functions within the group. These should be
listed and clearly defined in the SOP. The key definition
needed is for a computer system requiring validation; a
"validatable system". The practical determination of
validatable systems in day-to-day operations is the
responsibility of the operating committee.

Computer System Validation Management. The type of committees,
the definition of project managers (responsible users), and
their respective duties should be described in detail.

Validation Requirements. The SOP should also describe the
steps and responsibilities in the validation process. These
items can be incorporated in the validation protocol to ensure
compliance to the SOP and to ensure a complete documentation
package at the end of the process.

Validation Protocol

The easiest way to have consistency in the development of
validation protocols is to outline the requirements as a
checklist or a "fill in the blanks" document.

Documentation. Since the validation protocol is documentation
intensive, existing documents, reports, vendor manuals, etc.
should be used. The development of the protocol and the
methodology used for validation should fit the existing
management/committee structure whenever possible. Computer
systems validation should not create a new documentation
structure but rather pull together the necessary information
for documentation and testing from that which already exists.

Responsible People. The first part should list the computer
system and the person responsible for the validation process,
for example, the department head of the user group; the
responsible user. The other people responsible for the review,
implementation, and approval of the protocol should also be
listed.

Basis of Design. A Basis of Design/Basis of Operation section
should be included that can be used for both new and existing
systems. For new systems, this section will provide clarity
for purchase specifications. For an existing system it will
document information that probably does not exist elsewhere.
The main components of this section should include a narrative
description of what the computer system is intended to do, a
listing of requirements, the normal operating parameters
(current memory requirements, number of ports currently used,
etc.) and the absolute limits (maximum memory capacity, maximum
number of ports, etc.). It may also be helpful to identify
what the computer is not intended to do; this can prevent the
system from being overloaded or misused.

System Description. The exact system that is either currently
in operation or one that will be installed should be described.
The hardware and all peripherals should be listed along with
the applicable version of the operating software. The protocol
should make provision for the documentation that both of these
are certified at installation by the vendor using standard
diagnostic programs. Applications software needs to be
carefully documented and tested (verified) before it can be
loaded into the operating hardware for operational testing and
validation. The essential requirement for confidence in the
software verification process is assured by the meticulous
documentation of the specifications, planning, programming,
testing, debugging and final "test data" verifying testing
steps. Once the hardware/software information is collected,
then all of the other pertinent data concerning the interaction
with peripherals, equipment and instruments can be developed.

If a system is used for material control, the materials should
be adequately described (raw materials, package components,
work in process and/or finished products) along with the
methodology for switching to back up manual control.

Hardware/Software. Diagrams of the hardware and
hardware/software interactions are necessary for test plan
development and auditing of the validation process. Unless
your existing system is extremely well-controlled and
documented, these diagrams will probably be the first complete
identification of hardware/ software interactions.

Computer Room. Computer rooms are usually constructed
according to standard requirements of the major computer
manufacturers. The details of the particular specifications
for the computer room should be outlined for environmental
conditions (temperature, humidity, line voltage and radio
frequency interference). Consideration should be given to the
differences in requirements for large computer rooms with
multiple systems and for systems in laboratory areas. Each
system should have a set of operating manuals and historical
logs for: 1) hardware, 2) software, 3) critical events, 4)
back-ups, and 5) maintenance/downtime. These logs should be
maintained for periodic review and as an aid to change control.
Appropriate consideration should also be given to computer room
security.

Customer Acceptance. A new system installation and customer
acceptance should be formally documented. Any changes to the
system from the original specifications should be noted and all
related documentation, including diagrams, should be updated.
Existing systems should be documented as they currently are
installed.

SOP's. SOP's are necessary for all aspects of the operation,
maintenance, and change control of each system. They should be
coordinated between the various departments to be sure all
activities are covered. A grid of activities versus SOP's and
responsible departments incorporated in the protocol is an
effective doublecheck on procedures.

Training. The system operators and users need to be trained.
The responsible user should develop a training program in
conjunction with the operating committee and the
hardware/software suppliers, identify the operators/users and
train them. This training should be documented in the form of
a certificate for the individual and a training status log for
the system. All future training and system access must be
appropriately authorized and documented since this is the
cornerstone of internal system security and data integrity.

Operational Testing. The computer system should be
operationally tested by the operators/users. Operational
testing is the exercise of the verified applications software
in the certified hardware/operations software system using test
or simulated data. This can be accomplished in conjunction
with the documentation of the operational qualification of the
hardware/operating software and/or during the validation
testing.

Validation Testing. Validation testing is the exercise of the
verified applications software in a certified
hardware/operations software computer system using actual data
in a simulated mode or on line concurrent testing with real
time data. The requirements listed in the basis of
design/basis of operation part of the protocol are the
foundation for development of the test plan. The test plan
need not be absolutely perfect; the validation process is an
experimental study. If you find something wrong you have to
figure out how to fix it. The software verification process
should have eliminated the bugs but all possible circumstances
can not be foreseen. If the test plan is not complete, the
problems and solutions can be described in the summary report;
or, if they are serious problems, the solutions can be
incorporated in a new validation test plan.

Test Reports. The essential data for the test reports that
should be developed prior to testing are the
system/module/subsystem being tested; the tests to be
conducted; test references (if there are no literature
references, the committee members responsible for the test
design should be credited); test methodology, and acceptance
criteria.

Calibration. Prior to the initiation of testing, all
equipment, instruments and interconnects should be calibrated.

Testing. Testing should be carried out according to the
validation testing plan during realistic operating conditions.

Protocol Summary. A summary of the protocol documentation
including an analysis of the test results, the compliance audit
of the system, and any system modifications should be submitted
to the computer systems validation committee for their review
and approval. It is recommended that the system not be used
until final validation approval is received from the committee.

Audit Report. A report of an independent audit of the computer
validation process by an internal auditor (i.e. Quality
Assurance) should be included with the summary report to
management. The audit should compare the SOP and the initial
parts of the protocol (what the system should do) with the test
plan results (what the system actually does) and the summary
report conclusions.

Permanent File. The original validation documentation should be maintained by QAU since they are the regulatory contact concerning validation.

Identifying, Categorizing, and Prioritizing Computer Systems

The SOP and the protocol are the foundation of the validation process. The next step is to identify all the existing computer systems. One means of accomplishing this task is to send a survey form to all managers requesting information about computers used in their departments. Once this initial data is collected and analyzed, it is prudent to walk the buildings, room by room, blueprint in hand, to verify the information. When you are satisfied that you have an accurate listing of computers, the operating committee needs to categorize them into those requiring validation (validatable) and those not requiring validation (non-validatable). The types of decisions the committee will have to make concern the systems' impact on the production of pharmaceuticals and the manual back-ups to the computer systems; is the computer an "electronic file cabinet" used to store information for easy reference or is it used as the sole control of equipment, instruments and material? Now that the decision concerning the validation status has been made, a process of prioritization and resource allocation begins. The best approach is to select a small, easily documented and tested existing system. Choosing this type of system produces quick results and identifies problems in the administrative/mechanical part of the validation process. Once you iron out the difficulties with the first system, all the others will not be as difficult to validate.

Risk Analysis

A Risk Analysis should be filled out for all systems to aid in the determination of "validatable" systems and also to highlight the critical points for validation testing.

Project Tracking

All projects have a life of their own and each person has a different methodology and timing for getting the job done. Success in computer systems validation will depend on the operating committee's ability to keep the process moving. One means of project control is a tracking format that identifies the key steps in the validation process and the anticipated and actual completion dates.

This mechanism developed by the operating committee and filled out by the responsible user for all computer systems should be based on a standard outline of milestones, with additional key points customized to the particular type of computer system being validated. This is an outline of typical milestone dates: 1) submission of a validation assessment form; 2) committee review of the systems' need for validation; 3) development of the first draft protocol; 4) committee review of the draft protocol. (*Note:* steps 3 and 4 will be

repeated until the protocol is approved but a limit should be
set for the number of reviews.   If necessary, a review meeting
should be held to finalize the protocol); 5) preparation of
training manuals; 6) schedule of user training;
7) calibration of equipment/instruments prior to validation
testing; 8) validation testing schedule; 9) review of test
data and write summary report; 10) Quality Assurance audit of
validation process/documentation; 11) summary report review/
sign-off of validated system; 12) system use in operations.

Change Control

Changes in the computer system will inevitably take place from
the time the system is specified until it is installed,
validated and used in operations. These changes will usually
be captured in the validation documentation. Changes after
validation can alter the way the system is used and invalidate
the original validation work. Procedures (SOP's) for change
control need to be designed to keep a computer system operating
in a continuing state of control.

Periodic Review

The change control documentation for each computer system
should be reviewed periodically to ensure that no major change
nor a number of smaller changes have altered the function or
capability of the system. A good rule of thumb on timing is
not more than a year between reviews.

Revalidation

The FDA does not recognize the term revalidation. In their
lexicon, the protocol testing for any system is validation
whether or not a system has been previously validated.
Industry uses the term to define the continuing validation
testing of a previously validated system.

Summary

These are the essential steps in computer systems validation:

1. Recognition of the need for computer systems validation by
   management.
2. Formation of appropriate management committee(s).
3. Writing a procedure (SOP) for computer systems validation
   - what needs to be done.
4. Development of a working protocol outline - how to do it.
5. Identification of all the firm's computer systems.
6. Designating "responsible users" for all computer systems.
7. Determining which systems will be validated.
8. Drawing up a schedule for validating the computer systems
   on a priority basis.
9. Initiating the process by concentrating on only one
   computer system.
10. Annual reviews of the computer systems.

11. Steering Committee/Operating Committee monitoring of the
     administrative process for computer systems validation.

The validation of computer systems is an exercise in project
management. The fact that computers are involved does not mean
that the approach will be significantly different. It does
mean that the responsible users will be the focal point of
computer systems validation; they will have to assume the
responsibility for validation of the computer systems they use
just as they are responsible for all other compliance aspects
of their operation. MIS will play a key role since their
computer expertise and interface with the suppliers is required
throughout the process.

LITERATURE CITED

1. Guide to Inspection of Computerized Systems in Drug
    Processing, FDA Publication, February, 1983.
2. Motise, Paul J. Pharmaceutical Technology March, 1984,
    "Validation of Computerized Systems in the Control of Drug
    Processes: An FDA Perspective".
3. Motise, Paul J. Pharmaceutical Manufacturing July, 1984,
    "What to Expect When FDA Audits Computer Controlled
    Processes".
4. Kuzel, Norbert R. Pharmaceutical Technology September,
    1985, "Fundamentals of Computer System Validation and
    Documentation in the Pharmaceutical Industry".
5. PMA's Computer Systems Validation Commitee Pharmaceutical
    Technology May, 1986, "Validation Concepts for Computer
    Systems Used in the Manufacture of Drug Products".
6. "FDA Computer Investigation Expertise" and "Software
    Documentation" The Gold Sheet, July, 1986, 20, No. 7.
7. "Boehringer Ingelheim's Protocol" The Gold Sheet, August,
    1986, 20, No. 8.
8. Agalloco, J. Pharmaceutical Technology January, 1987,
    "Validation of Existing Computer Systems".
9. Kuzel, Norbert R. Pharmaceutical Technology February,
    1987, "Quality Assurance Auditing of Computer Systems".
10. Kahan, Jonathan S. MD&DI March, 1987, "Validating Computer
    Systems".

RECEIVED March 21, 1988

Chapter 12

Inspections and Final Report Audits for Environmental Studies

Patricia D. Royal

Battelle Ocean Sciences, 397 Washington Street, Duxbury, MA 02332

Enactment of the Good Laboratory Practices (GLP) regulation by the U.S. government was a direct result of an investigation conducted in 1975 by the U.S. Food and Drug Administration (FDA). That investigation evaluated the integrity of health assessment studies used to support registration of food additives, drugs and cosmetics under the Food, Drug and Cosmetic Act (FDCA). The conclusion of this investigation revealed flawed study conduct, inaccurate reporting, and inadequate data integrity. The FDA then implemented regulations affecting study conduct and data collection and retention. These regulations govern reporting requirements for health assessment studies used to support registration under FDCA and are commonly referred to as the GLPs (1).

Shortly thereafter, the U.S. Environmental Protection Agency (EPA) developed similar requirements under the Federal Insecticide, Fungicide, Rodenticide Act (FIFRA) and the Toxic Substances Control Act (TSCA) (2)(3). Together, these three regulations control the conduct and reporting requirements for all industrial safety assessment studies used to support registration of chemicals, pesticides, food additives, and drugs by EPA and FDA. Although environmental and chemical fate studies were included in the original TSCA/GLP, they were not specifically identified in FIFRA/GLPs. This inconsistency slowed the development and implementation of compliance programs by EPA to evaluate ecotox and the analytical chemistry associated with environmental studies. EPA is now proposing to fill that gap by redefining the scope of existing FIFRA/GLPs to include environmental and chemical fate studies.

Impact and Implementation

Expanding the scope of FIFRA/GLP to include environmental and chemical fate studies will take time. It will mean that studies described by EPA Hazard Evaluation Division, Office of Pesticide and Toxic Substances, for Environmental Fate and Residue Chemistry must meet the requirements outlined in the FIFRA/GLPs, and that

0097–6156/88/0369–0075$06.00/0
© 1988 American Chemical Society

laboratories and field operations conducting these studies will be
monitored by the EPA Office of Compliance Monitoring.
    It is always valuable when implementing a new program to assess
the impact of the program on the operations and the people involved.
One of the concerns voiced by scientists has been that GLPs interfere
with the advancement of science. This is especially true in field
testing programs and in analytical chemistry laboratories where
experimental procedural development often takes place.
Implementation of the GLPs into these programs will not dictate
scientific procedure, but will require putting scientific procedure
in writing, and will require documentation to trace the progress of
the study.
    Implementation at the testing site will mean developing SOPs
for application of test substance, randomization, location,
identification, and collection of samples, and establishing sample
custody procedures to ensure sample integrity while in transport to
the laboratory.
    In the laboratory, implementation will mean that the equipment
and the facility must be of adequate size to maintain the identity,
store, and analyze a variety of samples. Additional archival space
will be needed to provide storage for analytical samples, soil and
plant specimens. There must be procedures for documentation of
sample identification and receipt from the field to verify sample
custody procedures, proper sample storage and waste disposal. The
laboratory must have validated procedures for the type of chemical
analysis being conducted. Likewise, there must be documentation of
the training and analytical proficiency of the staff. Standard
Operating Procedures (SOPs) must be written to describe the
analytical methods used, and the maintenance and calibration of
equipment. Protocols or work plans must be established to specify
the objective of the study, personnel involved, equipment, and
methods, including criteria for accepting or rejecting data and the
frequency for running standards, spikes, and blanks, commonly known
as Quality Control (QC) standards. Applicable methods for developing
these programs can be found in 40 CFR, 136, and Quality Control in
Analytical Chemistry, by Kateman et al. (4)(5).
    The term Quality Assurance is often confused with Quality
Control. Quality Assurance is a program established to monitor
study conduct and reporting to ensure that they meet both external
and internal standards. In this regard, it is a management tool.
Quality Control (QC), on the other hand, is the criterion or
internal numerical standard on which the acceptability of data is
judged. By identifying some of the unique needs for verifying
conformance to standards in these studies, integration of the GLPs
into chemical fate and environmental studies will be easier. This
brings us to another impact of this new regulation, the development
of a Quality Assurance Unit (QAU) to monitor these studies.

The Quality Assurance Unit

The GLPs require the laboratory to establish an independent QAU to
monitor study conduct and audit the final report. This requirement
is needed to assure management and the government that the study is
being conducted according to the GLP regulations and that the

reported results accurately portray data collected for that study. Expanding existing regulations will require the formation of specialized Quality Assurance Units that can address the unique needs of environmental and chemical fate studies, and implement a program to monitor these studies. The purpose of the QAU is to ensure study integrity by monitoring these studies from the application of test substance, through collection of specimens or samples, to the chemical analysis, and to ensure the accuracy of data in the final report.

The GLPs also state that the QAU must inspect each critical phase of the study. In conducting inspections to assess the analytical chemistry phase of these studies, it is important for the QAU to identify how that phase fits into the overall study plan. This knowledge directs the inspection by determining what to look for and where to look. Basic questions that any QAU should ask in planning an effective inspection are given in Table I.

Table I. Planning an Effective Inspection

- Is this the beginning, middle, or end of the project?

- Is the purpose of the study to detect or to measure (qualitative vs. quantitative analysis)?

- What types of analyses are being conducted?

- What type of equipment is specified in the protocol or SOP?

- What are the calibration or QC requirements, including frequency, recovery, and control limits?

- What are the detection limits for the various analyses?

- What procedures are used to ensure sample custody and sample identification?

- Who is responsible for conducting the analyses and what is his or her training?

- Where are the SOPs kept and are they accessible to the staff at all times?

- Where are samples received and stored?

- Where are the data stored?

When conducting an assessment of an outside laboratory, the adequacy of the laboratory's QAU and its relationship to management should be determined in addition to assessing the laboratory operations. Questions directed toward evaluating the QAU might include those in Table II. It is often helpful fcr the QAU to make up a checklist; however, the list should be flexible and open-ended so that it can incorporate unanticipated events.

Table II. Evaluation of Laboratory Quality Assurance Unit

● Is there an independent QAU on site?

● To whom does the QAU report?

● What is the background of the QAU staff, and is the staff
  adequate to cover the amount and type of work being conducted?

● How does the QAU handle the Master Schedule? (A schedule
  required by the government specifying each study by chemical,
  type, and dates of conduct.)

● Does the QAU have SOPs describing inspection and auditing
  procedures?

On entering the laboratory, there are several rules of conduct
for inspectors that will help the QAU in conducting a successful
evaluation. Some Rules of Conduct are presented in Table III. Awareness of procedures used by inspectors will help the analytical
chemist to anticipate questions that might be asked by the inspector.

Table III. Rules of Conduct for Inspections

● As the inspector, you are there to observe and not to interfere
  or intimidate.

● Never interrupt a technician conducting a delicate procedure
  with a question that can wait until the procedure is complete.

● Be observant for the unexpected, either good or bad.

● Never assume; ask for the SOP, and check it to be sure that it
  is the same copy that is in the QAU. Follow the procedure in
  the SOP as it is being conducted.

● Review maintenance and calibration of equipment, placing special
  emphasis on QC acceptance criteria. This can be accomplished
  by reviewing control charts, percent recovery, parallel testing
  of new standards, and maintenance logs.

● Review the documentation of training for the staff. This can
  be done by reviewing curriculum vitae, job descriptions,
  proficiency testing records, education, and in-house training.

When conducting an inspection, several target areas must be
evaluated. Control limits or "charts" are helpful and should be
established by plotting the defined limits of acceptable quality
control. These charts are important tools for assessing laboratory
precision, accuracy, and reproducibility. They can be based on a
curve established from the high, mid, and low concentrations of a
standard analyte. Either the mid level or the average of the three
concentrations then becomes the mid-line for the control chart.
Acceptable levels of fluctuation for routine mid-level standards,

spikes, and blanks can then be identified and drawn onto the chart.
Control charts can also monitor percent recovery and reproducibility
or precision. The frequency suggested by the National Bureau of
Standards for control, mid-level standards, blanks, and spike runs
should equal about 5-10 percent of the sample load (6). Posted
control charts maintained daily can give substantial information,
including the following:

- Instant feedback to the technical staff on acceptable
  runs, drift, and out-of-control situations.

- A historical record of instrument operation.

- Verification of technical proficiency and variation
  between different staff.

Sample custody, more formally referred to as Chain-of-Custody
procedures, should be described in an SOP and reviewed. These
procedures are necessary to ensure sample integrity and identification from collection through transport to the laboratory, to
subsequent analysis and reporting. Various methods can be used from
hand-written sheets on which logging-in and out, storage, and
responsible personnel are indicated, to computerized bar code
setups, to more stringent systems in which sealed vials are used.
Whatever system is used, it should be adequate for the operation
and specified either in an SOP or a study-specific protocol or work
plan.

Another area needing close review during inspections is labeling and tracking of reagents and solutions. All reagents and
solutions should be reviewed to ensure their integrity, stability,
and proper labeling. Accountability, integrity, and stability can
be documented by establishing a reagent and solution log book. It
should indicate lot number, expiration date, storage requirements,
grade of material used, and disposal. Each reagent and solution
should be labeled to identify content, preparation date, expiration
date, storage requirements, and person who prepared the solution.

At the end of the inspection, it is helpful to hold a debriefing
with the Project Manager. This is important because it initiates
a dialogue and establishes a loop of communication between the QAU
and the Project Manager or Study Director. Misinterpretations can
be identified or additional data can be added to the report.
Suggestions for corrective action can be given in an informal way.
True GLP issues can be distinguished from scientific questions or
suggestions. Usually, the need for future inspections can be
discussed and a schedule determined.

The QAU must then write an Inspection Report and send it to the
Project Manager. The written Inspection Report should be complete
and objective. Suggested content is given in Table IV.

The Project Manager responds to the inspection report in
writing, identifying agreement or disagreement with the findings
and indicating corrective action; this closes the communication loop.
The completed report is then forwarded to the Director of the
Laboratory or other appropriate personnel to complete the monitoring
process or communication loop on a higher level.

GLPs also specify that the QAU must conduct an audit on the
final report. If the inspection phase has been conducted properly,

Table IV.   Characteristics of a Good Inspection Report

● It should stand on its own.

● It should identify the study, phase inspected, dates of
  inspection, items reviewed, and supporting data.

● It should clearly identify areas of compliance and noncompliance.

● It should identify those areas where improvement is suggested
  and methods for improvement.

● It should be worded such that the inspection procedure is a
  positive, useful experience for the laboratory.

● It should indicate the time of the next inspection, especially
  when corrective action has been indicated.

---

the audit should not be too time consuming. When conducting the
final report audit, the QAU must reconstruct the study to assure
that all the pieces are in place and that the study is complete.
There are several ways to verify study integrity; however, basically
all audits are divided into three parts.

   1.   The objective and scope of the study are determined
        by reviewing the protocol and relevant SOPs.

   2.   The raw data are reviewed for proper documentation
        and completeness.

   3.   The raw data are compared against the final report
        to ensure accurate presentation.

   This last part or phase is traditionally thought of as "the
audit." It can be accomplished in two basic ways: either using a
random number statistical approach or by a percentage or line
approach. In conducting an audit, it is important to remember that
some types of errors (usually the small ones) are random, for
example, a simple transcription error at the end of a long calculation, while others follow patterns and can have a cumulative impact,
such as an unacceptable calibration curve or even sample mix-up.
   At the end of the audit, a Final Report Narrative is written by
the QAU to the Project Manager. The format of the Narrative is
similar to that of the Inspection Report. Once again the Project
Manager responds in writing, thus establishing a loop; the completed
Narrative is then forwarded to upper management.
   The GLPs specify that the final report include a QA Statement
listing the dates of in-progress inspections, when they were sent
to the Study Director or Project Manager, and when they were sent to
upper management. This statement is to be signed by the QAU. The
QA Statement should not be confused with the GLP requirement for a
Compliance Statement. This statement verifies GLP compliance, and
is to be signed by the Study Director or Project Manager. Because
of potential confusion over these two statements, the Final Report
Narrative should address all areas of the report, and produce a
written dialogue between the Study Director, the Project Manager,

and the QAU. By addressing any problems in writing and by the
Study Director responding in writing, the distinction between these
two required statements can be clarified. The responsibility to
monitor the study and report on compliance is that of the QAU; the
responsibility to conduct the study in compliance with the
regulation is that of the Study Director.

Discussion

Enlarging the scope of the GLPs to include environmental and chemical
fate studies will have a substantial impact on the conduct of such
studies. Implementation of this regulation to include field testing
programs will bring new challenges to existing QAUs. However, by
defining the scope, the objectives, and responsibilities, and by
relying on past experience, we can begin to identify ways to meet
that responsibility.
      As the regulation becomes effective, it will be important for
EPA to explore the integration of the GLP regulation with existing
regulations commonly used for the Contract Laboratory Program (CLP).
The CLP program operated by EPA regulates laboratories conducting
chemical analyses under the Resource Conservation and Recovery Act
(RCRA) and Superfund programs (4)(7). They must identify overlap,
as well as differences, because many laboratories will be operating
under both regulations. Laboratories certified under the CLP
may think that they are in compliance with GLPs and not realize the
differences in the regulations. The CLP program is used to evaluate
laboratories contracted to analyze hazardous waste, while the
FIFRA/GLP program regulates the conduct of studies used to support
the registration of pesticides. Thus, it is conceivable that
some analyses may be regulated under both programs. Whereas the
CLP program specifies methodology and QC requirements, the GLP
regulation stresses record keeping and data accountability.
      In the past, the quality and integrity of environmental and
chemical fate studies have varied considerably. While I am not
recommending the development of a laboratory certification program
or mandatory methodology, the potential practical integration
of the CLP and GLP regulations could have a substantial impact on
the way environmental and chemical fate studies have been conducted.
Practical integration could improve the quality of these studies
in the future. Together, these regulatory programs could result in a
system to document methods and assess data integrity so that the
reliability of the results and conditions under which they were
produced could be verifiable in a way that would ensure accuracy,
reproducibility and successful legal review. The outcome of such a
system would enhance scientific acceptance, credibility, and public
confidence.

Summary

Applying GLP principles to field studies and analytical chemistry
operations will require identifying those operations that are
unique to the type of study and discipline. In this regard, the
importance of creatively adapting principles developed from in-house
monitoring situations to field study operations has been discussed.
The differences between QC and QA have been defined. Inspection

and audit procedures have been evaluated. The importance of establishing practical Chain-of-Custody procedures and Quality Control standards has been reviewed, as well as the importance of blending this regulation with existing regulations, and the role of the Quality Assurance Unit in monitoring these types of studies. The development of GLP compliance programs to monitor environmental studies programs is still in its infancy. Although the initial purpose of this regulation is to provide a mechanism for ensuring proper conduct and accurate reporting of data collected for environmental field testing programs used to support registration of pesticides and chemicals, its potential application will undoubtedly go beyond these activities. We are already seeing GLP record keeping principles being applied to municipal and industrial pollution monitoring programs, Environmental Impact Statements, RCRA/Superfund operations and court review. Basic principles of good record keeping and documentation are fundamental to this regulation and to good science, and are therefore universal in their application.

Literature Cited

1. The Good Laboratory Practices, FDA, 1987, FR 43, No. 247.
2. The Good Laboratory Practices, FIFRA/EPA, 40 CFR, Part 160.
3. The Good Laboratory Practices, TSCA/EPA, 40 CFR, Part 792.
4. Guidelines Establishing Test Procedures for the Analysis of Pollutants Under the Clean Water Act, 40 CFR, Part 136.
5. Kateman, G.; Pijpers, F.W. Quality Control in Analytical Chemistry; John Wiley & Son: New York, 1981.
6. Principles of Quality Assurance of Chemical Measurements, U.S. Department of Commerce, 1985, NBS-SIR 85-3105.
7. Identification and Listing of Hazardous Waste, 40 CFR, Part 261.

RECEIVED January 29, 1988

Chapter 13

Quality Assurance in Analytical Laboratories

An EPA Perspective

Willa Y. Garner

U.S. Environmental Protection Agency, EN-342,
Washington, DC 20460

The Federal Insecticide, Fungicide and Rodenticide Act
(FIFRA) Good Laboratory Practice (GLP) Standards regulations (1) are intended to ensure that regulatory studies
are conducted with good planning and execution, complete
documentation and validation, and integrity. Official
GLP inspections include a review and evaluation of the
testing facilities as well as an audit of the data generated by those facilities. Chemistry auditors evaluate
the entire study for environmental, residue, product
chemistry and metabolism studies, but only the analytical
phases of health effects and ecotoxicology studies.
Sample collection, handling, transfer, and storage procedures are steps in an analytical study that may offer
an opportunity for loss of sample integrity and must be
documented in detail. Registrants are responsible for the
retention of their raw data which must be maintained as
long as the registration, which it supports, is active.

The conduct of a chemistry-related good laboratory practice (GLP)
laboratory inspection and data audit will be discussed in this
paper. This will be accomplished by describing the basic audit
procedure, then digressing into the objectives of an audit and the
primary problem areas that have been experienced.

Before addressing the fundamentals of an audit, let us review
some of the regulatory background and history of the GLP regulations.
These days, as you know, the regulatory testing laboratory has a new
partner, the Federal auditor or inspector, who will be critically
reviewing all aspects of the selected study as well as those of the
ongoing operations. This person is a verifier of accounts, as the
dictionary phrases it. He is sent to verify that the public's trust
in science is well founded.

The regulated community is fully aware that the Federal presence
is the result of revelations that some laboratories were submitting
false or faulty data as the basis for obtaining permits to sell

their products. There were inconsistencies between the raw data and
the final reports. The test protocols were poorly written and the
test data were not properly maintained. Consequently, in granting
permission to use toxic chemicals to control agricultural pests,
Congress required its public servants to assure themselves that no
harm would occur to the users of these products or to the environment when used according to label directions.

As a result, we find ourselves in a position, as Federal
regulatory officials, to insist that there will be quality assurance
and quality control as an inherent accompaniment of analytical work
and that analytical data accompanying all regulatory submissions
shall be carried out in compliance with good laboratory practice
regulations. The GLP regulations are intended to ensure that
studies are conducted with good planning and execution, complete
documentation and validation, and integrity.

Proposed Generic GLP Regulations

An EPA committee worked for over a year on formulating a set of
generic GLP regulations to cover all TSCA and FIFRA regulatory
studies. The efforts of this committee were published in the Federal
Register (2) on December 28, 1987, for a 90-day comment period. The
proposed FIFRA GLP regulations appear as the last chapter in this
volume. When the FIFRA GLP regulations become final they will cover
not only health effects studies, but also environmental fate, residue, metabolism, ecological effects, and efficacy studies. Field
studies will be covered as well as laboratory work. For studies
started before the GLP regulations become final rule, and completed
after that date, the portion of the study conducted after the final
rule date must have been conducted under the GLP regulations with
proper documentation as to which part of the study was conducted
under GLP and which part was not. It is anticipated that the Revised
FIFRA GLP regulations will become final in the summer of 1988.

U.S.E.P.A. GLP Inspections

The laboratory to be inspected will receive a letter approximately
two weeks before the Agency inspection team arrives that specifies
which studies will be audited and if a laboratory GLP inspection is
to be included. Upon arrival, the inspector will present official
credentials and a Notice of Inspection form. The GLP portion of
the audit is now conducted as if GLPs for all types of studies were
in effect. For those laboratories conducting non-GLP studies, this
is done to give an idea of what to expect when, and if, the GLP
regulations become law. The laboratory inspection aspects will be
reviewed briefly and then the data audit portion will be discussed.
Many of you have expressed an interest in a format for your
master schedule. Figure 1 depicts the format Mobay Chemical Corp.
uses. It is self explanatory and covers the items required in the
GLP regulations (test substance; test system; nature of study; study
initiation date; current status; sponsor identity, if applicable;
and name of study director). For a contract laboratory, the sponsor's identity must appear on the master schedule sheet for each
study listed. There are several terms that require definition. In

the proposed GLPs, experimental start date means the first date
the test substance is applied to the test system, and the experimental termination date is the last date on which data are collected
directly from the study. These dates must appear in the protocol.
The study initiation date, which is the date that is entered on the
master schedule, is defined as the date the protocol is signed by
the study director. The study completion date will refer to the
date that the final report is signed by the study director.

The laboratory area is inspected to ascertain if space and
equipment are adequate for the size of the staff and the scheduled
workload. All equipment, such as gas and liquid chromatographs,
infra red spectrometers, nuclear magnetic resonance spectrometers,
etc., have service, preventative maintenance, or calibration logs.
Laboratory requirements differ as to the amount of information documented in their maintenance logs. Some laboratories provide little
information and others provide extensive amounts. Figure 2 depicts
the information that Analytical Development Corporation (ADC) records
for their gas chromatograph usage. Balance and pH meters must have
calibration logs and must be calibrated and/or standardized either
once daily or prior to use, whichever is appropriate.

Dry chemicals, solvents and stock solutions must be properly
labeled. The labels used at Tegeris Laboratories (Figure 3) give an
example of the items to be addressed. Each storage container for a
test, control, or reference substance must be labeled by name, CAS
or code number, batch number, and expiration date, if appropriate.
Where appropriate, storage conditions necessary to maintain the
identity, strength, purity, and composition of these substances must
be given. For studies of more than four weeks' duration, reserve
samples from each batch of test, control, and reference substance
must be retained as long as the quality of the preparation affords
evaluation. All radioactive materials must be labeled as such.

All equipment, including balances and hoods, must be regularly
maintained and so documented. Minimally, hoods should be checked on
a yearly schedule. Dow Chemical Company uses the label shown in
Figure 4 to document hood maintenance checks. Refrigerators and
freezers must have a temperature recorder of some type or be manually
checked and the temperatures recorded. This covers the major GLP
related items in the analytical laboratory.

Standard Operating Procedures (SOPs) are also an important
concept of GLP regulations. If followed, they ensure that a laboratory's compliance with GLP regulations is well defined and consistent, regardless of the personnel conducting the research. SOPs
must be developed for such topics as: specifying the operation,
calibration, and maintenance of pieces of equipment; defining how to
record raw data and what raw data to record; explaining what information is to be logged when chemicals are received; indicating how
to design studies and take samples in the laboratory or in the field;
and explaining how to input and verify computerized data.

Data Audits

A data audit may be either priority or routine. Priority audits are
conducted if a discrepancy, data gap, or other potential violation
is suspected. Routine audits of studies submitted for pesticide

MOBAY CHEMICAL CORPORATION
CORPORATE TOXICOLOGY DEPARTMENT
STANLEY RESEARCH CENTER, STILWELL, KANSAS

*B = BATCH #
F = FORMULA #
L = LOT #
R = REFERENCE #

STUDIES IN PROGRESS

PAGE 1

SCHEDULE AS OF SEPTEMBER 1986

COMPOUND (B,F,L,&R)*	TYPE OF STUDY	SPECIES	REQUEST	LOCATION EXPER/PROC/PATH	INITIATION DATE	EXPECTED COMPLETION

Figure 1.　Master schedule format (Reproduced with permission from Mobay Chemical Co.).

DATE	NAME	PROJ.	COLUMN	COLUMN			MAKE-UP		ml/min		PURGE SYSTEM	
			SIZE COATING	TYPE	HEAD PSI	(ml/min) FLOW	TYPE	TOTAL FLOW	SPLIT VENT	PURGE VENT	ON	OFF

TEMPERATURES			SIGNAL	RANGE ATTEN.	RECORDER SPEED PER MIN.	RAMP INFO.	SAMPLE INFORMATION					GENERAL MAINTENANCE
OVEN	INLET	DET					TYPE	#	VOL	AUTO	SOLVENT	SEPTUM, GLASSWOOL, PROBLEMS

Figure 2.　Information collected on gas chromatograph maintenance log (Reproduced with permission from Analytical Development Corp.).

DATE RECEIVED: _____

EXPIRATION DATE: _____

INITIALS: _____

COMPOUND: _____

SOLVENT: _____ CONC.: _____

STORAGE: _____ EXP. DATE: _____

PREPARER'S INITIALS: _____

PREPARATION DATE: _____

Figure 3.　Information labels for dry chemicals and solvents (top) and stock solutions (bottom) (Reproduced with permission from Tegeris Laboratories).

registration are carried out at a facility approximately at 15-month intervals. The interval may be shorter for audits of certain pivotal data submitted for reregistration or for the development of a registration standard.

The chemistry auditor usually audits only the analytical chemistry portions for health effects or ecotoxicology studies and the entire data file for environmental, residue, product chemistry, and metabolism studies.

Health Effects and Ecotoxicology Studies

For the health effects studies, the dosage preparations, including test substance and reference standard characterization and stability, and the diet preparations are reviewed by the auditor. Diet preparation aspects include homogeneity of the test chemical in the diet and the stability of this material in the diet covering the period from the time it is mixed through the feeding period. The auditor will also ascertain if the protocol was followed. If a change in the study design occurs prior to the event, the protocol should be formally amended to cover it. Any protocol deviations noted during the study should be adequately documented. It is important that the protocol approval date precede the experimental starting date. The same issues are addressed for the chemistry portions of the ecotoxicology studies. Feed and water data, including analyses for nutrients, contaminants, and other pertinent parameters will also be reviewed by the chemistry auditor. Clinical chemistry is another area subject to review during the chemistry audit.

There are many sources of variability related to the sampling, handling, transfer, and preservation of samples. The preparation, sampling, and analysis of animal feeds deserve special attention. It is an established fact that the difficulties of distributing parts per thousand, parts per million, and even parts per billion of a test substance homogeneously into a feed mixture are monumental.

In looking at the dosage form of the test article, the dosage preparation method is evaluated and the calculations for the concentration levels are checked. Proof of stability of the test article during the period of the study and the analytical procedures used to test for stability are evaluated. Proof of homogeneity, stability, and proper concentration of the test material in the diet and the analytical procedures used to ascertain homogeneity and stability are also evaluated. These properties must be addressed prior to the initiation of the study. In most cases, the concentration of the test substance in the carrier is expected to be within + 10% of nominal for concentrations greater than 10 ppm in the diet, if experienced analysts are utilizing validated specific methods. If this limit cannot be met, the protocol should be amended to show why this was not possible, and why this would not impact upon the validity of the study.

Included is a graph (Figure 5) from an article by William Horwitz which relates analytical precision to concentration. It shows that the analytical variability increases as the concentration decreases. The Horwitz data were generated from collaborative studies where methodology was exactly defined. The data should be

INDUSTRIAL HYGIENE HOOD SURVEY

HOOD IDENTIFICATION _____

DATE OF SURVEY _____

SURVEYOR _____

STATIC PRESSURES*:
    SASH CLOSED: _____ INCHES WATER
    SASH FULLY OPEN: _____ INCHES WATER

FACE VELOCITIES, FPM @ SASH OPENINGS, INCHES:
_____ @ _____
_____ @ _____
_____ @ _____

CHEMICAL USED IN EVALUATING HOOD _____

RESTRICTIONS:
_____
_____
_____
_____

*NOTE: IF STATIC PRESSURE READINGS VARY 25% FROM THOSE MEASURED, THE
EXHAUST SYSTEM SHOULD BE CHECKED.

Figure 4. Information collected to document hood
maintenance (Reproduced with permission from
Dow Chemical Co.).

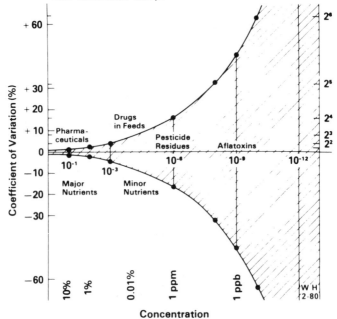

Concentration

Figure 5. Graph relating analytical precision to concentration. (Reproduced from Ref. 3. Copy-
right 1981 American Chemical Society.)

repeatable by a single analyst consistantly using the same exact method. An easily remembered reference point is that at 1 ppm in the diet, the coefficient of variation is $\pm$ 16%.

Records for documentation of the mixing procedure used to achieve homogeneity of the test substance in the carrier must be available for audit. Prior to the analysis of the study samples, all analytical procedures must be validated in terms of recovery, reproducibility, sensitivity, freedom from interference, and accuracy.

If the test substance mixture is shown to be unstable in the diet, it is important to either prepare the test substance-carrier mixture more frequently to achieve stability or show unequivocally that the decrease in concentration is due to the chemical binding to the carrier and that it would still be biologically available to the test animal, i.e., that it would not be volatilizing or decomposing into other compounds.

Environmental Fate, Residue, and Metabolism Studies

For the pristine chemistry studies which include studies such as hydrolysis, soil and water photolysis, soil dissipation, and rotational crop under environmental fate, metabolism studies, residue studies, and product chemistry studies, such as vapor pressure, octanol-water partition coefficient, and water solubility, the total study is audited. This includes the GLP issues, such as adherence to protocols, SOPs, and record accountability; completeness of raw data; the validation of data points; and the overall scientific issues.

The chemical aspects of these studies focus primarily on the chemical characterization of the test substance and/or mixture. The identity of the test chemical should be proven, and the analytical procedures used, such as gas or liquid chromatography, nuclear magnetic resonance spectrometry, or mass spectroscopy, should be available for audit. This would include the chromatograms or spectra from these analyses. It is imperative that raw data be left intact as they emerge from an instrument to maintain data integrity. Chromatographic printouts are to remain attached and in sequence. If some data points are not used in the final report, the reason is to be documented and those not used are to remain with the study file. No raw data are to be discarded.

To comply with the portion of EPA Pesticide Registration Notice 86-5, which states that oversize computer printouts or fold-out pages not be included in the registration package, it is suggested that photocopies be made of the chromatograms, and that the photocopies be cut to fit on an 8 1/2 x 11 inch page. Column conditions and other chromatographic parameters must appear in the raw data. Types of information to be documented are given in Figures 6 and 7 for gas and liquid chromatography, respectively. Quality control during sample analyses is an important aspect in the conduct of a scientifically sound study. Chemistry auditors will ascertain if replicates, recoveries, and reagent blanks were assayed with the samples, if an independent audit mixture was employed to check out proper machine functioning prior to use, and if the slope sensitivity was set correctly to assure proper integration for GC and HPLC analyses.

Operator _____	Date _____
Stationary Phase _____	Instrument _____ Detector _____
Film Thickness _____	Range _____ Attenuation _____
Column No. _____ Type _____	Flow Rates, cc/min.
Length _____ OD ____ ID ____	Make-up _____ Type _____
Carrier Gas _____	Hydrogen _____ Air _____
U _____ Flow _____	On Column ☐ Split ☐ Splitless Injection ☐
Chart Speed _____	Ratio _____ Hole Time _____
Sample _____	Temperature-Det. _____ Inj. _____
Size _____ Solvent _____	Column Initial _____ Time _____
Concentrations _____	Rate _____ Final _____ Time _____

Figure 6.   Gas chromatographic parameters documentation.
(Reproduced with permission from Supelco, Inc.)

DATE _____ OPERATOR _____ CHROMAT. NO. _____

COL. NO. _____ LENGTH _____ ID _____ packed with _____ PCT

(PHASE) _____ on (SUPPORT) _____

MOBILE PHASE AND GRADIENT _____

_____

TEMPS. (Reservoir) _____ (Col.) _____ (Det.) _____

PRESSURE _____ FLOW RATE _____ mi./min.

DET. _____ SENS. _____ CHART
SPEED _____ min.
sec./in.

SAMPLE _____

SAMPLE CONC. _____ INJ. AMT. _____

Figure 7.   Liquid chromatographic parameters documentation.
(Reproduced with permission from Anspec Co.)

Test Substance/Mixture Characterization

The method of test substance synthesis or its source should be made
a part of the documentation. This would apply to any test chemical,
whether it is a technical material, a formulation, a metabolite, a
by-product, or a radiolabeled compound. Any impurities greater than
0.1 % in the test material should be identified and quantified. If
a commercial or technical lot is specified for the study, comparison
should be made between the test chemical and its commercial counter-
part. The test substance, or mixture, should meet routine specifications for chemical composition and physical properties. The source
and lot or batch number of the test article and any diluants, such
as acetone or corn oil, should be given in the raw data. Since, in
almost all cases, the test substance, or mixture, will be shipped to
the laboratory performing the study, a bill of lading describing the
test material as to name, purity, lot number, quantity shipped,
handling procedures, etc., is needed along with chemical receipt
records to provide a complete paper trail to prove the transfer,
handling, and receipt of the test material. Storage and custodial
procedures at the test facility are necessary documentation for each
test substance. Auditors will ask to see the archived sample of the
test substance for studies whose term exceeds four weeks.

At this point, it should be stressed that when characterization
of the identity of parent chemical and/or metabolites is required in
a study, that identity must be confirmed by an alternate technique.
Data reported without application of suitable confirmatory techniques may not only be worthless, but what is worse, incorrect
information may be seriously misleading and may be unrectifiable.

All data points should be used; one should not be selective,
i.e., one from column A and another from column B! Use statistical
tests to determine if data points in the set are truely outliers.

One expects biological data to be full of perturbations resulting from the many outside influences on the particular property
we are measuring. Consequently, we get zig-zag patterns of these
properties with time, complete with standard errors extending from
each point which often do not overlap one another. An auditor should
really begin to worry about the quality of the observations when
there is no reasonable variability component. Less than usual variability suggests that some averaging has been going on. One can
average out quite a few wild results, if they are in opposite directions, and get a fairly decent mean. If one takes enough widely
variable data points one can hide poor data by this method.

A bulk test chemical inventory must be maintained for labeled
and unlabeled test materials (Figure 8) which describes the chemical
as to name, appearance, quantity, lot number, storage conditions,
etc. The rest of the form would have columns for date, person re-
moving the material from stock, the quantity taken, the quantity
remaining, and a column for the person receiving the material to
sign for it. The purity of the test chemical must be shown prior to
the initiation of the study, as well as its stability throughout the
study. The analytical procedures used to assure stability must also
be available for audit.

Reference standards must be characterized as to purity, batch
or lot number, source, storage requirements, and traceability, and

ABC FORM NO.: 307

COMPOUND _____   STRUCTURE/
SPECIFIC ACTIVITY _____   FORMULA _____

TOTAL ACTIVITY _____   RADIOISOTOPE_____

SOURCE _____   PURITY _____

DATE RECEIVED _____   LOT NO. _____

RADIATION SAFETY OFFICER _____ INVESTIGATOR _____

DATE	uCi USED	uCi REMAINING	LOCATION OF WORKING SOLUTION	NAME & INITIALS INVESTIGATOR	STUDY #

Figure 8. Information captured for the bulk test chemical usage inventory (Reproduced with permission from Analytical Bio-Chemistry Laboratories, Inc.).

have periodic purity assays if the same lot is used over an extended
period of time.

Sample Collection and Handling - Field Studies

Sample collection, handling, and storage are steps in an analytical
study that offer many opportunities for loss of integrity of the
sample and must be described in full detail. Good judgement cannot
be assumed; details must be provided. Complete sample control must
be maintained from the time the samples are taken in the field,
if this is the case, through their analysis in the laboratory to
final storage. Figure 9 depicts the type of information required
for Dow Chemical Company residue field trials. Tools used to acquire
the samples must be described, as well as the sample containers. It
is a known fact that bottle cap liners and aluminum foil which may be
coated with drawing oil may also be sources of contamination. These
aspects must be considered when planning sample collections. One
needs to describe how the sample containers are cleaned and how the
samples will be shipped and then stored when they arrive at the
laboratory prior to analysis, as well as the temperature and length
of time for storage. Exposure to light and air are important considerations. Storage stability data must be provided for the
same matrix and cover the time period that samples are stored prior
to analysis. All samples must be logged in and assigned unique
numbers which are fully traceable. Fragile samples will not need to
be retained beyond quality assurance review. To ascertain sample
storage and handling procedures, the chemistry auditor often sets up
the situation of "I am a sample arriving at the laboratory. What
are your procedures for handling me from my point of arrival through
extraction and analysis to final storage?"

For all of the studies we audit, we ask for a curriculum vitae
on each of the staff members who are conducting and/or are involved
with the study. We want to know about their education, experience,
and training in the area they are working.

Data Validation

After having looked through the laboratory's files for all of the
information we have discussed, the auditor now begins the analytical data validation phase of the audit. Usually, approximately
10% of the data points appearing in the report submitted to EPA are
randomly selected and validated. This means tracing all the raw
data involved in obtaining the selected data point in the report
back to their initiation. Sometimes, the audit of a study will be
from photocopies rather than from the original records. To document
that the photocopy is a "true" copy, it must be certified. Rohm
and Haas Company uses the stamps depicted in Figure 10 on their
photocopies to assure validity.

In looking through the raw data, the auditor also checks for
overwrites and incorrectly executed cross-throughs, as depicted in
Figure 11. Overwrites and use of white-out are prohibited, according to the GLP regulations, and cross-throughs should be executed as
shown, with the person's initials, the date executed, and the reason
for the change. Frequently, insufficient space is available for

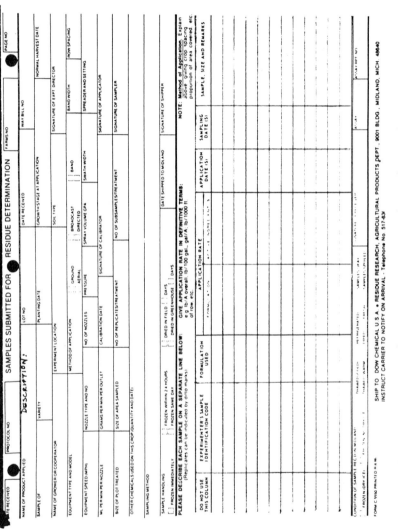

Figure 9. Information collected for residue field trials (Reproduced with permission from Dow Chemical Co.).

This is an exact copy of
the original document.
BY _____ DATE 20 May 87

This is an exact copy
reduced in size of the
original document.
BY _____ DATE 14 May 87

Figure 10.   Rhetoric for documentation of a photocopy as
a "true" copy (Reproduced with permission
from Rohm and Haas Co.).

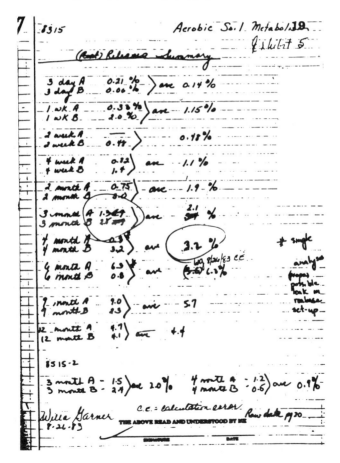

Figure 11.   Sample laboratory notebook page depicting
overwrites and incorrectly and correctly
executed cross-throughs (reproduced with
permission from Uniroyal Chemical Co., Inc.).

describing the reason for the change. To conserve space, speed up the correction process, and provide consistency throughout the laboratory, ADC developed the following numerical listing:

EXPLANATION OF NOTEBOOK ENTRY ERRORS

1. Misspelled
2. Mathematical Error
3. Wrong Entry (date, sample no., word, etc.)
4. Transposition or Sequencing Error
5. Transcription Error
6. Procedural Change
7. Wrong Conclusion
8. Illegible Entry
9. Unnecessary Entry
10. Footnoted Explanation
11. Additional Comment

Each time an error is made, it is initialed, dated, and one of the code numbers in the list is placed next to the initials and circled. A copy of the list is placed in the front of each notebook for reference. Pencil or white out are not to be used under any circumstances. Note at the bottom of Figure 11 the place for the witness or supervisor to sign. During audits, we have had many discussions about this. The consensus in the Office of Compliance Monitoring is, if there is a place for a signature, sign it. If this practice is not acceptable to the laboratory, an SOP should be developed to explain this deviation in the use of the form.

Raw data used to be a very simple concept: they were the numbers actually indicated by a measuring device, whether it was the sum of weights on a balance, a determination on an instrument dial, or a measurement on a recorder chart. The analyst had full control and responsibility over the production of the data at every step. With mechanization and automation, where the responsibility for instrument calibration is assigned to the manufacturer of the equipment or the proper functioning of the instruments is assumed to be built-in by the instrument designers and computer operators, the production of data has shifted from a straight line function, entirely under the direct supervision of the professional scientist, to a more complicated operation managed by a laboratory director. Automated instruments measure the samples, execute the manipulations, determine the response, perform the calculations, and present the final answer in whatever form or units desired. The final value may be copied from a dial, recorded on tape, drawn on a chart, or not presented at all, to be stored in a computer for coordination with past and future values, presenting the entire sequence as the result of the experiment. These final results are raw data just as much as the direct measurements are. Whether the results come directly from manual observations or from automated instruments is not important. What we should be asking is "Are these data correct?", "Are they original data?", and "How do we know?" It is sometimes difficult to reconstruct computer generated data points because of dilution factors, rounding of numbers, etc. Check to see if you can recalculate the numbers before you have to do it for an auditor.

Data transformation steps should be documented in laboratory notebooks.

Inspection Closing

At the end of the facilities inspection and data audit, the inspector
will present the laboratory with a Receipt for Samples form. This
form lists all of the copies of documents, samples, etc., the inspection team collected for use in documenting the findings of the
audit in their report. The laboratory will be given a closing session in which the auditors and the Inspector will discuss their
findings. Frequently, this conference also provides a time for a
question and answer session or an exchange of ideas.

It is 10:00 p.m. Do you know where your raw data are?
It is very important that you do. It could be costly if you do not.
Under Section 8 of FIFRA, the registrants are responsible for their
raw data, for its integrity, and for its protection.

The Code of Federal Regulations, 40 CFR 14, §169.2 (k), Maintenance of Records, states that "Records containing research data
relating to registered pesticides, including all test reports submitted to the Agency in support of a tolerance petition, all underlying raw data, and interpretations and evaluations thereof, whether
in the possession of the producer or in the possession of the
independent testing facility or laboratory (if any) which performed such tests on behalf of the producer. These records shall be
retained as long as the registration is valid and the producer is in
business."

Under the paragraph entitled Civil Penalties in Section 14 of
FIFRA, "Any registrant, commercial applicator, wholesaler, dealer,
retailer, or other distributor who violates any provision of this
Act may be assessed a civil penalty by the Administrator of not
more than $5,000 for each offense."

In assessing the results from these audits, for the most part,
the lack of raw data has been the most critical deficiency, along
with occasional findings of careless science. If data are missing,
a civil fine may be levied, and the study may have to be repeated.
An auditor or inspector's responsibility is to present and document
the facts: They do not invalidate studies, and they do not levy
fines or penalties.

The level of sophistication shown in the implementation of the
GLP regulations varies greatly between the different laboratories.
Most contract laboratories are well into compliance since they
perform FDA related GLP studies and have been involved with QA for
several years now. Second in rank come the in-house company laboratories who also perform studies for FDA. There has been some
information exchange between the toxicology groups and the analytical
groups. The rest of the companies, especially those units who do
only environmental or residue chemistry studies, are for the most
part behind their counterparts, and many still have a long way to
go to catch up.

The GLP regulations are being accepted as the minimum standards
of research quality; however, compliance with the principles outlined
in the GLPs does not in itself ensure quality research data. Any

research quality assurance program should include the GLP concepts as part of its basic structure. It cannot be overemphasized that an effective quality assurance program must have the support and involvement of multiple levels of management and research personnel.

Acknowledgments

Appreciation is extended to Dean Hill, USEPA, NEIC, Denver, CO, for his helpful comments and suggestions in the review of this paper.

Literature Cited

1. Pesticide Programs; Good Laboratory Practice Standards; Final Rule, Fed. Reg., 48:53946, November 29, 1983.
2. Federal Insecticide, Fungicide and Rodenticide Act (FIFRA) and Toxic Substances Control Act (TSCA); Good Laboratory Practice Standards; Proposed Rules, Fed. Reg., 52:48920, December 28, 1987.
3. Horwitz, W. Analytical Measurements: How Do You Know Your Results Are Right? In "The Pesticide Chemist and Modern Toxicologist" (1981) S. Kris Bandal, Gino Marco, Leon Goldberg, Marguerite Leng, Eds.; ACS Symposium Series 160, American Chemical Society, Washington, DC.

RECEIVED March 21, 1988

Chapter 14

Quality Assurance for a Field Trials Program

Testing Residues of Agricultural Chemicals

James P. Ussary

Agricultural Products Group, Eastern Research Center—Goldsboro,
ICI Americas, Goldsboro, NC 27530

This paper will describe a GLP compliance program developed to deal with the field phase of a complex residue
trials program and will describe how the field and
laboratory phases of the studies are integrated. An
agricultural chemical residue trials program is a special problem to a quality assurance unit trying to implement the EPA Good Laboratory Practices (GLP) regulations. The field trials are often dispersed over a
vast geographic area, are located in remote areas, each
trial is done by a different person, and the trial activities are affected by such things as weather, local
agricultural practices, and seasonal differences. The
laboratory portions of the residue studies are less
troublesome for the quality assurance unit, but the
laboratory and field portions of the study must be integrated so that there is good communication between
the field and the laboratory personnel, and there must
be an easily followed continuity of the records from
the field to the laboratory for the quality assurance
program to be effective.

An effective GLP compliance program is a disciplined way to document
scientific studies that, if done properly, is the total ambience in
which the studies are done. For a GLP compliance program to be effective, it must be supported by the management of the organization
and be a discipline that permeates the entire organization. At ICI
management considers the GLP compliance program to be important and
cost effective. This program has become a routine part of the day-
to-day activities of the organization. It affects the work of each
of the organization's employees. The GLP discipline has been developed by management and the Quality Assurance Section so that the
compliance program has been accepted as necessary, if not desirable,
and it has been supported and enhanced by those that come under its
regulation.

0097–6156/88/0369–0099$06.00/0
© 1988 American Chemical Society

At ICI Americas Inc. Eastern Research Center - Goldsboro, a
formal GLP compliance program was begun in 1979. The initial program was focused on the laboratory functions, but as the laboratory
program matured the study directors and Quality Assurance personnel
began to require additional documentation for the field portions of
the studies so that both the laboratory and field portions of these
studies are now done according to the principles of the EPA Good
Laboratory Practices regulations.

The purpose of this paper is to describe the GLP compliance
system developed for the Residue Chemistry Field Trials Program at
ICI Americas Inc. This program covers crop residue and
environmental fate studies.

Education

Education of the scientist whose work is subject to the GLP regulations as well as other members of the organization is critical to
the success of the GLP program. Everybody needs to understand the
purpose of the program as well as its requirements. The program
will be enhanced by the people that have this understanding. The
administrative staff and other non-technical people will support the
program, and the technical people will help develop the program so
it will be useful to themselves and still meet the regulatory requirements. All new scientific employees who will be working on
studies that are subject to the GLP compliance program are required
to have a training conference with a member of the Quality Assurance
Section staff. Special sessions are held for the summer college
student employees. Seminars are held for Marketing and Technical
Service employees to explain the program. Discussions are held at
company technical meetings to exchange ideas.

Quality Assurance Section representatives periodically attend
staff meetings of each section that does work subject to the GLP
regulations to discuss quality assurance concerns. It has been much
easier to get compliance by discussing problems than to simply make
recommendations in audit reports.

Study Management

The study management must be organized so that there is good communication between the laboratory and the field personnel. One person, the study director, must be in control of the study and be
aware of what is happening with that study at all times. At ICI the
study director is always the scientist who will write the final report. This is usually the Residue Chemistry analytical team leader
who is responsible for writing the protocol and whose team will analyze the samples. All major decisions about the trial are made by
the study director. For complex studies such as soil dissipation
studies, crop rotation studies, and leaching studies, a study coordinator may be appointed to help write the study protocol and actually supervise the field activities. The coordinator would be a
scientist with expertise in the type of study being conducted. This
person is responsible for keeping the study director advised about
the progress of the study.

The field portions of the trials are done by the Research Farm
staffs or the Development Section technical representatives. The
Development technical representatives are highly trained biologists
who are strategically located throughout the United States and are
responsible for conducting efficacy and residue trials. Most use
their home as a base of operation and work alone. The Developmental
Chemistry Section is responsible for sample processing and analyses.
Although there are at least two, and often three, R&D sections involved with each study, it is clearly understood that any problems
or questions concerning a study are directed to the study director
or the study coordinator.

Developing the Protocol

It is required that before any scientific work is started for a
study that there be a protocol approved by the study director, the
study director's manager, and that the protocol be audited by
Quality Assurance.

Study protocols are written by the study director in cooperation with the Development Section regional managers, the Research
Farms managers, and the Registration manager. Each of these people
provide necessary information for the protocol design. Each study
is assigned a unique protocol number and each trial within the study
is assigned a unique number. These numbers identify the study
throughout the field and laboratory phases of the study.

When the first draft of a protocol is typed, it is sent to the
Quality Assurance Section to be audited. The protocol is checked
for those details required by the Good Laboratory Practices guidelines. Comments about each protocol are sent to the study director.
When the final version of the protocol is typed, it is again sent to
Quality Assurance for review. The original copy of the signed and
dated protocol is filed in the archive.

Deviations in the conduct of the study from the protocol must
be properly documented. There must be a formal protocol amendment
signed and dated by the study director for any prospective change in
the conduct of the study. This includes changing such things as the
location of the trial, the application rate, or the formulation of
the test chemical. Unavoidable changes such as those caused by
adverse weather, seasonal variations, or wildlife damage must be
clearly documented in the raw data and a written opinion by the
study director about the impact of each change on the study must be
put in the study file.

The Master Schedule

When a protocol is issued, the study is put on the Master Schedule.
The Master Schedule is a computer-generated document that can be
formatted and sorted to accommodate the needs of various participants in the study as well as the Quality Assurance Section. It
contains the protocol number, the trial numbers, the study title,
the proposed start and finish dates of field and analytical segments
of the study, the proposed reporting date, the names of field participants, the name of the study director, and other information that
may be useful to the users of the master schedule. As a study

progresses, the proposed dates are updated to real dates. The
master schedule is updated and reprinted monthly. Quality Assurance
uses the Master Schedule to develop an audit and inspection
schedule.

Standard Operating Procedures

Standard operating procedures (SOPs) are required for all routine
activities that are critical to the successful outcome of the study
including quality assurance procedures and inspections. Most of the
SOPs for the field activities are written by the field scientists
with guidance from the Quality Assurance Section. For the field
activities, it is required that at each site there be SOPs for such
things as how field plots are established and the plot boundaries
marked, the maintenance of sample freezers, how to calibrate and
maintain chemical balances and chemical application equipment, and
how to obtain test chemicals. If an SOP for a critical item of
equipment is not available, the study would be considered to be not
in compliance with the principles of the GLP guidelines. The distribution of SOPs is carefully controlled so that when an SOP is
changed, all outdated copies can be exchanged for the new version.

The Test Chemical

Accurate and complete records must be maintained on the chemical
product used for the trials. Records are kept on the product from
the time the technical chemical is received until it is applied to
the test plot. The Large Scale Formulation Laboratory, where small
batches of product for testing are manufactured, is operated according to GLP guidelines. There are records kept on the receipt of all
product ingredients, the time and method of making the product, container sizes, and shipping information. Each batch of product is
assigned a unique batch number. Records are kept of how the formulated product was made. This includes all weights, machine settings, and other details that would be needed to reproduce the batch
of product. When the product leaves the Large Scale Formulation
Laboratory, it is sent to the Shipping and Receiving department.
Large Scale Formulation Laboratory is provided a signed and dated
receipt by Shipping and Receiving.

Requests for product by the field investigators are sent to the
sample coordinator. The request specifies that the product is to be
used in a residue trial. The sample coordinator maintains a list of
batches that are acceptable to the study directors. These are usually batches that have been made with fully characterized technical
chemical. The sample coordinator sends an order which specifies the
batch number and container size to Shipping and Receiving. Container sizes from 1/2 pint to 1 gallon are available and are provided
according to the trial's needs. The product is sent to the field
investigator along with a 2-copy packing list. One of these copies
must be signed and dated when the product is received and returned
to Shipping and Receiving. It is then placed in the permanent
Archive. If this receipt is not received within 21 days, a followup
letter is sent. The second copy of the packing list is for the
field investigator's records. When the product is used, the batch

number and the visual appearance of the product are recorded in the trial records.

A small sample of each batch of technical chemical and formulated product is stored in the chemical archive.

Residue Samples

An easily followed record of a residue sample from the time it is harvested until it is analyzed is a necessary part of a GLP compliance program. This can be difficult when the possession of the sample may change several times between the field and the analytical laboratory, and a sample may be renumbered at one or more of these stopovers. Each field investigator is issued a block of unique numbers for residue samples. A number is affixed to the sample when it is harvested and identifies that sample throughout its existence including the results from that sample in the final report. For some studies, the sample numbers are assigned in the protocol.

Records are kept of when a sample was collected, the method of collection, who collected the sample, what the elapsed time was between harvest and freezing, the conditions under which it was stored, how it was shipped to the laboratory, and when it was shipped. When a sample arrives at the laboratory, the condition of the sample is checked and recorded. Then the information on the sample bag (sample number, application rate, preharvest interval, etc.) is compared to the trial information sheets which are submitted along with the samples. Any omissions or discrepancies are corrected at that time. If there is an omission or discrepancy that cannot be easily corrected by a telephone call from the sample processing laboratory to the field scientist, the study director is notified. The study director must make the decision about the validity of the sample and put a note in the data file explaining how the problem was corrected or that the problem could not be corrected and the trial is to be abandoned. If the trial is dropped, Quality Assurance is notified, the trial is deleted from the active Master Schedule, and an explanation is put in the study file.

The samples are logged into the Laboratory Sample Inventory System, then processed for analysis. Records are kept on the method of processing, the technician who did the processing, and the storage location in the freezer.

When the samples are to be analyzed they are requested in writing by an analytical team leader (usually the study director). When that person takes possession of the samples, it is noted in the inventory system as well as when the samples are returned. While the samples are in the laboratory, the times and dates they are removed and returned to the laboratory freezer are recorded in the laboratory data sheets.

The Sample Storage Freezers

Sample storage freezers located at the Research Farms and those of the Residue laboratories are considered to be limited access archives and are kept locked. Access is limited to the sample preparation laboratory employees. All movements of samples in and out of the freezers are recorded. One freezer in the Residue Laboratory

is maintained as a free access "working" freezer for general use by
the analysts. Samples are kept in this freezer only while they are
being analyzed.

The Residue Laboratory freezers are all equipped with temperature alarms and emergency power generators and are covered by a service contract. Each freezer has a 7-day thermograph. The calibration of each thermograph is checked against a mercury thermometer
each time the chart is changed. The freezers are also equipped with
an electronic monitoring system that is programmed to give an oral
message by telephone to certain extensions at the site and to certain employees' homes if any freezer malfunctions. During non-
working hours the temperature of each freezer is monitored hourly by
security guards, and as added security a technician telephones the
speaking monitoring system each night before retiring and records
the temperature of each freezer in a log book.

In the field, all types of freezer storage facilities are used.
The investigator may have a food freezer at his home that is used
for residue samples or the freezer may be in a rented mini-storage
warehouse across the city from his home. The samples may be stored
in a commercial freezer facility, in a walk-in freezer at a university, or some other facility chosen by the field investigator. Each
freezer is equipped with a recording thermometer. Each time the
chart is changed, the calibration is checked with a mercury thermometer and the reading recorded. All freezer records are retained in
the data archive.

Data Reporting

Data forms were designed to satisfy the requirements of EPA Standard
Evaluation Procedures and the Good Laboratory Practice guidelines.
Seventeen field data forms were designed, most of which are used for
every field trial. These forms include everything from a signature
page to a page for miscellaneous observations. There is also a one-
page form for the study director to indicate which of the seventeen
data forms should be completed by the field scientist. There is a
postcard form that is sent to the study director by the field scientist when a trial is initiated. A second postcard is sent if a
trial fails. These postcards are circulated to the Quality
Assurance Section. All forms are completed according to an SOP.

These data forms were designed by a committee of personnel from
the Residue Laboratory, Data Processing, Quality Assurance, and the
field scientists that would be required to use the forms. This committee had a significant impact on the GLP compliance program because it brought all the people together that had a direct interest
in the data. The result was a set of data forms that, contrary to
most predictions, has been accepted, used properly, and received
favorable comments from the field.

Data Security

The field data, as well as the laboratory data, must be secured from
loss or tampering. In the field, the data forms when not in use are
kept in the investigator's office files. At the Research Farms
where there are several employees, these files are kept locked and

access is limited to specified individuals. The completed data
forms are sent to the laboratory along with the samples. The original copies are filed in a locked fire-proof filing cabinet in the
Sample Processing Laboratory. When the analysis of the samples is
started, the field data forms are transferred to the study director
who keeps them in a locked file when not in use. The study director
signs a receipt for the data. All laboratory data for in-progress
trials are also kept in a locked file when not in use. When the report has been completed, all of the field and laboratory data and
the original copy of the final report are stored in a limited access
archive.

Auditing in the Field

An Agricultural Chemical Field Trials program is a special problem
to a Quality Assurance Section. A study often has 12 to 15 trials
scattered over a large portion of the US. An agricultural chemicals
company may have 100 or more such studies each year. It has been
estimated that to inspect every trial when critical phases were
being done, as is done in laboratory studies, at least 25 inspectors
would be needed to cover the US. Each would need a car, office,
telephone, and travel expenses. They would be quite busy through
the summer but there would be very little for them to do during the
cold months. As an achievable alternative, the Quality Assurance
Section inspects the techniques and records used by individual field
investigators rather than concentrating on the details of each study. By watching an investigator perform a function in one trial, it
is assumed that all trials done by that person will be done similarly. While visiting a field site, all of the trials being done in
that general vicinity are visited. The plot markers are checked to
see if they match the plot diagram and other information in the
data. A subjective evaluation of each trial is made about the general appearance of the plots, site security, potential for the plots
to be disturbed by other work in the area, etc.
    The chemical storage area is checked to be certain that the
product batches recorded in the data are actually on hand and
properly stored.
    The equipment used for chemical application and sampling is in-
spected. This is usually a casual inspection. If the equipment is
clean, the hoses and belts appear to be in good condition, and
equipment is stored properly, the investigator probably takes proper
care of his equipment. The maintenance records of the equipment are
checked.
    The field investigator's residue sample handling procedures and
equipment are always inspected. It is policy that samples be held
at the field locations for only short periods; however, accurate records are kept on the storage and handling conditions from the time
a sample is collected until it is shipped to the laboratory. The
elapsed time from collection until a sample is placed in a freezer
must be recorded as well as the location of the freezer and the
temperature of the freezer during the storage. The sample storage
freezer and the records kept on it are always inspected.
    The field investigator's office is always visited. Trials data
should be neatly and securely stored. Also, the data forms for each

trial are inspected and they should be current, signed, dated, etc.
There should be a complete set of current standard operating procedures that pertain to that investigator's operation. If the investigator has an assistant, training records and a job description
should be on file for that person. (Training records and job descriptions for scientific personnel located at the Research Farms
are maintained at the farm sites. Training records and job descriptions for Development technical representatives are maintained at
the home office.) At each site there should be equipment maintenance records. At each of the Research farms, there should be a
current organization chart that shows how the conduct of the trials
is managed.

After a field inspection is completed, any deficiencies found
or recommendations for improvement are discussed with the investigator. This is followed by a written report. The investigator then
may append any comments he may wish to make to the report. He signs
and dates the report then sends it to his supervisor who signs and
dates the report and returns it to the Quality Assurance auditor.
The report is filed in the confidential Quality Assurance files.
Any needed follow-ups on deficiencies are usually done by telephone.

It is the objective of the Quality Assurance Section to inspect
each Research Farm and each Development scientist annually; however,
the inspections are often more frequent at sites where complex non-
routine studies are being done such as pond studies, soil dissipation studies, or groundwater studies.

Summary

An effective GLP compliance program for field agricultural chemicals
studies has been developed by ICI Americas Inc. The development of
this program was mandated and supported by management. The implementation of the program was achieved by involving the field
scientists in its design and the development of the procedures.

Proper organization of the study management is necessary for an
effective quality assurance program. Each study is directed by the
scientist who will write the final report but may be assisted by a
coordinator. The coordinator also usually assists the study director in designing the study. All questions about the trial during
its conduct in the field go to the study director or the
coordinator.

The trial records, as well as being complete, correct, and
usable, must be easily traced throughout the trial. The data forms,
standard operating procedures, test substance records, etc. were designed by the people that would use them with input from the Quality
Assurance Section, when needed.

An inspection program for field trials was developed that provides reasonable assurance that the studies are being conducted according to the protocol and the GLP regulations. It was deemed im-
practical to inspect each trial so the procedures used by the field
scientists are inspected and it is assumed that all of the trials
done by that trialist are conducted similarly. At least annual
inspections are made at each site except for complex studies where
multiple inspections are usually made.

RECEIVED January 29, 1988

Chapter 15

Quality Assurance in Contract Laboratories

Commitment to Excellence

Charles R. Ganz and Kathleen H. Faltynski

EN-CAS Analytical Laboratories, 2359 Farrington Point Drive, Winston-Salem, NC 27107

The present paper discusses one contract
laboratory's approach to setting up and operating a
successful quality assurance (QA) program. The
discussion focuses on the QA philosophy of the
laboratory, the ingredients included in the QA
program to help make it viable, and the
responsibilities which both sponsor and contract
laboratory must accept in order to optimize the
contracting relationship and produce quality
studies.

The main product of the contract analytical laboratory is numbers.
A number is an abstract entity. Unlike more tangible items, its
quality cannot be estimated by conventional means such as taking it
for a test drive or plugging it into an electrical outlet to see if
it operates. In order to evaluate the quality of a numerical
result, a sponsor must look beyond the number itself to the
laboratory which generated it; to its people, its integrity ... in
short, its commitment to excellence.

What Constitutes Excellence

What constitutes excellence? In rather simple terms, excellence in
the contract laboratory can be defined as producing work which
consistently meets high standards of quality. Emphasis is placed
on consistency since without this factor the sponsor's confidence
and trust in the laboratory will quickly evaporate. In more
specific terms, a contract laboratory committed to quality work
should perform its work in a way which builds and maintains mutual
confidence between sponsor and laboratory. At a minimum, the
laboratory should strive to ensure that the work being done
satisfies the sponsor's study objectives and is produced promptly
and accurately. In addition, frequent communication between the
contract laboratory and the sponsor helps to assure the sponsor
that adequate progress is being made and that problems are being

0097-6156/88/0369-0107$06.00/0
© 1988 American Chemical Society

promptly addressed. Likewise, the presence of well-documented results and well-organized study files gives the sponsor a strong indication that their study is being properly controlled and managed.

Finally, the contract lab should make a concerted effort to assess the scientific validity and reasonableness of the analytical data before reporting it to the sponsor. Reporting of obvious outliers can quickly erode a sponsor's confidence in the laboratory's data review process and thus in the study's results. There is a growing tendency on the part of analytical laboratories to accept results generated by sophisticated instruments and computers as inherently correct. All too often conceptual and manipulative errors (misplaced decimal points, transposition errors, forgotten dilutions, etc.) are hidden beneath the prima facie value calculated by the computer. The laboratory's study director and data reviewers, before giving final approval to a study, should look at the reasonableness of the results in terms of expected trends or results. Findings such as significantly higher or lower than anticipated residues, or inverse relationships between application rates or time-after-application and residues found, should prompt the laboratory to inspect the samples, sample history, and the analytical data to ensure that no procedural errors are evident. This may involve reanalyzing selected samples to obtain confirming results. If no errors are found then a discussion of the findings with the sponsor may lead to a possible explanation for the inconsistent data. Some common occurrences which we have found responsible for inconsistent results include labeling errors, accidental interchange of samples during collection or processing, non-representative samples, lab or field contamination, unusual binding of analyte to substrate and incorrect communication of active ingredient content by the sponsor.

The above listing suggests that the informational and mechanical, as well as the scientific aspects of the work must all be optimized to ensure results of consistent quality.

What Makes Excellence Happen

The achievement of excellence requires a concerted effort and commitment at all levels of the organization. Management must initiate this effort by making it clear to all personnel that quality is the overriding objective of the organization. (Although, in a capitalist system, profit is said to be the primary objective of a commercial enterprise, in our experience, quality and profit seem to go hand in hand). Management's commitment must give more than just lip service to the idea of quality. Its commitment must be backed up by actions that demonstrate to the staff that management is willing to pay the price of rejecting and reworking results that do not measure up to the laboratory's quality standards. The commitment equation is complete only when the laboratory's staff fully accept the importance of the quality objective to the success of the organization.

To ensure the growth of the quality objective, management must resist both internal and external pressures which might subvert its

goals. Such pressures might include staff resistance and
unrealistic deadlines. Once the climate for achieving excellence
is created by management, then the introduction of a well
thought-out quality assurance (QA) program can serve as a framework
for producing quality work.

Objectives and Strategies Used in Formulating the QA Program

During the very early stages of formulating our QA program, we set
as our underlying goal the creation of a program which would not
only meet government-mandated good laboratory practice (GLP)
requirements, but would also serve as a framework for
organizational excellence.
     In developing the strategies to be used in setting up the QA
program, we concluded that, to be effective, the program needed to
be practical and achievable. In other words, it needed to be
designed to minimize impedance of the work flow and, at the same
time, be relatively easy for lab personnel to learn and implement.
In addition, the program needed to be flexible enough to allow for
the exercise of scientific judgement and creativity and to allow
the laboratory to respond to real-world situations where time
requirements and study variables could not all be predicted ahead
of time.
     Our efforts to design a program which satisfied the above
requirements began with an exhaustive review of every aspect of our
work flow. Every potential step, from the initial contact by the
sponsor through archiving the final report and preparing for a
possible client or agency audit, was identified and critically
evaluated. Before accepting a step as a necessary operation to be
addressed in the laboratory's QA program, we asked ourselves the
following questions:
     - Is the operation necessary?
     - Who should be responsible?
     - What detailed procedures should be followed?
     - Do the procedures allow for scientific judgement?
     - Are the procedures realistic and practical?
     - What documentation will be needed?
     - Who and what files should receive copies of the
       documentation?
     - And, finally, is there a simpler way to accomplish the same
       result?

     With a detailed flow chart of laboratory operations in hand,
we set out to design a QA manual which would not simply contain a
series of standard operating procedures (SOP's) but would, in
addition, serve as a training and reference manual for producing
quality work in the laboratory. The manual would establish
performance standards as well as specify procedures for monitoring
the quality of laboratory operations, for correcting operational
deficiencies, and for instituting improved operating procedures.
     The topics included in the EN-CAS QA manual are listed in
Table I. We have found that these major divisions form a logical
framework into which the individual SOP's can be inserted.

TABLE I.   STANDARD OPERATING PROCEDURES CLASSIFICATIONS

SECTION	TITLE
I.0	GENERAL REQUIREMENTS
II.0	SAMPLE HANDLING AND TRACKING
III.0	LABORATORY OPERATING PROCEDURES
IV.0	ANALYTICAL METHODS
V.0	DATA HANDLING AND DOCUMENTATION PROCEDURES
VI.0	RECORD KEEPING AND ARCHIVES
VII.0	REPORT WRITING AND APPROVAL
VIII.0	QUALITY ASSURANCE
IX.0	SEQUENCE OF EVENTS FLOWCHART (STUDY SEGMENTS AND DOCUMENTATION)

As can be seen in Table II, the subtopics in the Laboratory
Operations section include both very specific SOP's for instrument
operation as well as a number of generic SOP's which serve as
training guides for conducting analytical studies.

To aid the analyst in providing necessary documentation, forms
were developed which prompted analysts to enter the needed
information. One such form is illustrated in Figure 1. The top
sections of the "Standards Preparation Sheet" shown in the figure
direct the analyst to provide a wide range of information deemed
essential for maintaining an adequate audit trail. In addition,
the information in the upper right hand corner instructs the
analyst on the proper distribution of the multiple copies of the
form.

Our final strategy for implementing a successful quality
assurance program involved the selection of a QA Officer (QAO) who
would be capable of carrying out both the letter and the spirit of
the program. With our holistic approach to achieving excellence,
we set as one of our selection criteria that the person have the
requisite education and experience to permit an in-depth
understanding of both the mechanical and the scientific aspects of
laboratory operations. To help ensure that the program would be
managed in a practical manner, the successful candidate for the QAO
position was assigned to work as an analyst in our laboratories for
six months prior to assuming the QA Officer's responsibilities.
This "hands-on" experience provided the incumbent with an essential
perspective on what may or may not be realistic to expect in
laboratory operations.

Table II. DETAILS OF SECTION III OF QA MANUAL

SOP Number	Revision Number	Title
III-1	0	GENERAL LABORATORY PROCEDURES APPLICABLE TO ALL STUDIES
III-2	0	HANDLING BULK CHEMICALS
III-3	0	PREPARING STANDARDS AND REAGENTS
III-4	0	PREPARING AND CLEANING GLASSWARE
III-5	0	PREPARING SAMPLES FOR ANALYSIS
III-6	0	EXTRACTING ANALYTES
III-7	0	CLEAN-UP PROCEDURES FOR EXTRACTS
III-8	0	CONCENTRATION PROCEDURES FOR EXTRACTS
III-9	0	GENERAL PRACTICES FOR USING AND MAINTAINING MEASUREMENT INSTRUMENTS (SEE SEPARATE PAGE FOR LISTINGS OF SOP's FOR INDIVIDUAL INSTRUMENTS)
III-10	0	QUALITY CONTROL PROCEDURES

```
            EN-CAS LABORATORIES              IWhite Copy --- Notebook
                                             IPink Copy  --- Std Registry(QA)
         STANDARD PREPARATION SHEET          IYellow Copy --- Job File
         -------------------------           I(Use Xerox Copies of White Copy
                                             I To Cross Reference Multiple
         Part 1 - Stock Solutions            I Analytes in Stds Registry)
                                             I_____
```

Analyst   _____   _____
                  Name   ./   Signature   Date
Bal Mod/Ser. # _____   Notebook Ref. _____

                        (NOM)   (FOUND)
Balance Check Wt. 1 _____   _____ Job _____

Balance Check Wt. 2 _____   _____

         SUPER STOCK IDENTIFYING INFORMATION

                    Analyte 1   Analyte 2   Analyte 3   Analyte 4

Name
E Number
Source
Batch Code
% Purity
Appearance
Expiration Date

         SUPER STOCK PREPARATION

Tare +
Tare
Wt or Vol Used
Dec. % Purity
Wt. or Vol.
 Active Ingred.
Solvent ID
Vol. Solv. Used
Concentration

         MIXED STOCK PREPARATION (if necessary)

Vol Super Stock Used (1) _____ml + (2) _____ml + (3) _____ml + (4) _____ml

    in total volume of _____ml of _____(solvent)

Conc. of Mixed Stock (1) ____ug/ml + (2) ____ug/ml + (3)   ____ug/ml + (4) ____ml

         COMMENTS (i.e., difficultly in dissolving, etc.)

_____

_____

         Figure 1. Standard Preparation Sheet

Since the QAO is, by definition, placed in the position of critically evaluating results generated by the scientific staff, we felt it essential that the QAO have the requisite maturity and diplomatic skill to earn the support and respect of the staff. The QAO could then use these attributes to convert a naturally adversarial relationship to a cooperative one and thus provide the springboard for positive change. Consequently, the QAO needed to be tough-minded enough to enforce the QA requirements but open-minded enough to allow for an exchange of ideas regarding QA operations.

From management's perspective, the QAO needed to be capable of critically evaluating laboratory operations, recognizing the need for changes and improvements, and recommending practical alternatives. In turn, the QAO would need no less than full management support in order to properly execute the QA unit's role as the purveyor of excellence. We, therefore, deemed it prudent and necessary to have the QAO report directly to company management.

Functions Performed by the QAU

We have identified five functional areas in which our QA unit should operate. These areas include inspection, reporting, record-keeping, custodial and advisory. The main tasks performed in each functional area can be summarized as follows:

Inspection

- Reviewing all raw data and analytical results.
- Reviewing final study reports.
- Inspecting notebooks, use-logs, facilities, lab operations and analyst's laboratory practices.
- Performing, upon request by the sponsor, on-site inspections for selected field studies.
- Participating in lab audits performed by sponsors and agencies.

Reporting

- Issuing inspection reports to the study director and management.
- Issuing signed QA inspection statements for all study reports.

Record Keeping

- Receiving copies of all study protocols.
- Maintaining a master list of studies.
- Updating and maintaining the QA manual.
- Maintaining a registry documenting the preparation of analytical standard solutions.
- Maintaining a file of curricula vitae for all lab personnel.
- Overseeing the archives of completed study files.

Training

- Orienting new employees to QA procedures.
- Training employees in new and modified QA procedures.
- Refresher training in existing QA procedures.

Custodial

- Custodian of analytical reference standards, receiving and cataloging incoming standards, discarding expired standards.
- Insuring that shared laboratory areas are kept organized and maintained.

Advisory

- Advising management about general QA problems and needs, recommending corrective actions and improvements.
- Based on inspections, feeding back to analysts and study directors where and how to improve their QA practices.

The Sponsor's Role in Producing Quality Work

As hard as a contract laboratory may try, it cannot generate quality studies from its effort alone. The sponsor must be a partner in the pursuit of excellence. Drawing from our experience, we have targeted several areas in which sufficient support from our sponsors is frequently lacking. Additional sponsor attention to providing assistance in the areas cited would almost certainly increase the likelihood of producing higher quality study results. Some of these tasks may be delegated by the sponsor to the contract laboratory. However, such assignments must be clearly specified by the sponsor at the outset of the study.

The sponsor needs to provide a clear idea of what the study will actually involve. Vague descriptions of the study requirements make proper planning virtually impossible. The study protocol needs to include a detailed analytical section written either by the sponsor or by the contract lab after consultation with the analytical department of the sponsor. This increases the likelihood of anticipating potential problems. One or more contact persons should be designated by the sponsor to supply additional information about the study, the test material, and analytical methods as needs arise during the study. Sufficient lead time should be given to the contract lab both to gain an understanding of the analytical requirements of the study and to make provision for realistic scheduling.

Prior to a field study, enough untreated control material should be provided to allow the lab to develop and validate adequate analytical methods. The control material should match the test samples as closely as possible to minimize the matrix variations which might affect the performance of the method. Development and validation of a method using a matrix which does not closely resemble the actual test matrix frequently results in a method which is not adequate for the actual study samples. The method revisions required in such a case represent a clear waste of

time and money. Sufficient control material should also be
provided along with the actual study samples so that an adequate
number of procedural recoveries can be run during the study.
Naturally, properly certified analytical standards are required, as
well as all the descriptive information about the standards which
is required by GLP regulations.

The sponsor or contract facility responsible for the field
work needs to organize the sample packaging and shipping so that
sample losses due to breakage and cross-contamination are
minimized. Packaging samples in logical groupings also reduces
sample handling and greatly assists the contract laboratory in
promptly cataloging samples so that sample integrity is not
compromised and missing samples can be easily spotted.

It is important to include, in all shipments, a shipping list
with a logical set of sample codes, as well as a key to their
interpretation. The shipping lists, if properly designed and
certified, can serve as a transfer of custody form. If the
laboratory is given adequate forewarning when samples are to be
shipped, it can make provision for storing the samples when they
arrive or for swiftly initiating tracing procedures when samples
are not received at the expected time.

There are several other actions, which, in our experience,
have proven extremely useful in helping build quality into studies.
We have, for example, found that a small pre-study, performed prior
to an actual study, will frequently uncover a majority of the
logistics problems likely to be encountered in the main study.
This approach may not be needed for routine studies but can often
make the difference between success and failure in a more complex
study. We have similarly found that having a member of our QA or
technical staff on site during the critical first few days of a
field study often allows potential sampling and sample handling
problems to be spotted and rectified before the quality and
integrity of the study is compromised. Finally, the practice of
fortifying known amounts of test material into untreated controls
at the test site is strongly encouraged. These samples, when
shipped, stored and analyzed alongside the actual study samples,
provide a good indication that sample integrity has been maintained
and analytical methodology is in control.

What a Sponsor Should Expect from the Contract Lab

The sponsor, having met its obligations in properly laying the
groundwork for a quality study should expect the contract
laboratory to assume its proper role in assuring that study quality
is maintained.

At a minimum, the sponsor should expect the contract lab to
consistently produce accurate results and to provide these results
to the sponsor in a timely manner. The contract lab needs to
display honesty and candor in their scheduling estimates and in
reporting problems which may arise from time to time. This is
especially true if the problem is the result of an error on the
contract laboratory's part. This will be easier if the laboratory
has made an effort to keep the lines of communication open and
active so that the sponsor is well aware of the progress (or lack

thereof) of the study. As part of this communication process, the contract laboratory should be capable and willing to provide technical advice regarding the analytical aspects of the sponsor's study. Finally, the laboratory needs to maintain documentation and files such that data reviews by the sponsor can be easily accomplished. The achievement of all of these goals can be greatly facilitated by having in place a QA program which both meets EPA requirements and is designed and operated to ensure the laboratory's continuing commitment to producing quality work.

Some Costs and Benefits of a Well-Operated QA Program

Let's look at the balance sheet of costs and benefits for a well-operated QA program.

On the cost side, we certainly can expect an increase in overhead expense since personnel and office space will be needed to run the program. Similarly, paperwork will increase as documentation requirements expand. QA requirements will also reduce the contract laboratory's ability to provide quick response and turnaround time in rush and emergency situations. Report generation will likewise be slowed. It is also probable that some staff objections will be raised. Finally, if managers and QA inspectors are doing their jobs, there is likely to be an initial increase in the number of work packages returned to analysts for either further documentation or reanalysis until staff members learn to work under the new requirements. It is our contention and experience that QA costs can be controlled and minimized if management makes a strong effort to anticipate potential problems at the outset and designs the QA program to address these problems.

What return can management expect from its investment in a well-operated QA program? Firstly, the QA program should produce an objective, hopefully unbiased, review and inspection system for the laboratory's operations. This should almost certainly increase management's confidence in the data being generated by the laboratory. It is difficult to compute a dollar value for an intangible attribute such as confidence. However, having it is likely to allow many managers (and sponsors) to sleep better every night. Secondly, if, as we have surmised in formulating our QA program, the program is to serve partly as a training vehicle then one result should be a better-trained staff, making fewer procedural errors, and producing better science. Thirdly, the documentation requirements of a good QA program should result in better organized data. This in turn should reduce the time the staff needs to spend on data review and report generation as well as the time required identifying and uncovering the causes of problems. Additionally, if the QA program is properly designed and operated, the laboratory should be in compliance with government-mandated GLP requirements and thus should encounter relatively few problems in supporting their results during agency and client audits.

Finally, the ultimate benefit of a QA program which fosters a laboratory's commitment to excellence should be the production of high quality, scientific studies upon which reasoned, regulatory decisions can be based.

RECEIVED January 29, 1988

Chapter 16

University Response to Good Laboratory Practices

Case History

Terry D. Spittler

Analytical Laboratories, Cornell University, New York State
Agricultural Experiment Station, Geneva, NY 14456-0462

The Analytical Laboratories consist of twenty-two
chemists in a variety of research activities including several Federally funded programs potentially
answerable to GLP's. To assess the impending procedural and fiscal impact of GLP adherence on the
Laboratories programs, a review of all phases was
conducted at our request by EPA to establish the
current level of compliance. A detailed response is
presented, addressing both those tenets which may be
met by procedural adjustment, and those mandates,
particularly with regard to facilities and personnel,
that resources and/or university policy will not
permit meeting. Alternatives are discussed.

University participation in programs that are regulatory in nature,
or that are driven by regulatory procedures, occupy unique nitches
in their respective academic communities. While certainly no
operational description would be all encompassing, a few germane
generalizations will help focus this discussion relative to Good
Laboratory Practices. Very few (read that as "NO", with an escape
clause) university entities exist solely as regulatory units, that
is, as facilities, staff and program devoted to a particular
statutory objective, answerable only to Agency personnel and super-
vised entirely by its mandates. Instead, oversight plus technical
and administrative direction is usually the responsibility of a
university faculty member, assigned by the dean, who receives no
direct salary support from the program -- the % Faculty Year
Equivalent is considered to be a contribution by the university.
The activities within a program are a combination of research and
routine determinations, with output subject to the discipline and
peer review guidelines of the academic department in which a
regulatory contingent resides. Finally, physical boundaries are
indistinct at a university, with much space and equipment being for
common use when and as needed. Thus, a regulatory unit existing

0097-6156/88/0369-0117$06.00/0
© 1988 American Chemical Society

within a university jurisdiction may be strictly defined in time
and dollars, but very loosely defined in space.
        The nature of regulatory-directed studies is also such that,
compared to basic research, they are frequently held in low esteem
by administrators and discipline peers alike. As unjust as this
may sound, it is a fact of academic life and those of us who are
engaged in these areas must live and work with it -- some of it is
justified. There is a certain repetitiveness to these studies that
obscures their scientific merit and the original research that may
have preceded these phases. Unfortunately, it is this mundane
aspect that is remembered when space and resource allocations are
made. Any mandates that reinforce this image of hackneyed data
production will only hurt the position of practical and/or
regulatory programs.
        All of the above allude to the point that regulatory par-
ticipation must coexist with the academic and basic research
components of a university. Hosted programs may be independent in
funding, operation and philosophy, but under no circumstances will
they be allowed to infringe upon the academic and research functions. The simple expedient of having the directors' salaries
university (rather than program) derived assures that decisions can
be made without undue influence.

THE CORNELL LABORATORY

This next segment pertains to the subject facility, the Analytical
Laboratories, housed in the Food Science Department, College of
Life Sciences, a statutory unit of Cornell University and the State
University of New York. It is located at the New York State
Agricultural Experiment Station, Geneva, NY. Twenty-two chemists,
plus support personnel, are engaged in a variety of research,
regulatory and contract endeavors, including:
  a. a regulatory contract with the Divisions of Food Inspection
      and Plant Industries for analysis of compliance and complaint
      sample components of feeds and fertilizers sold under the
      labeling jurisdiction of the New York State Department of
      Agriculture and Markets.
  b. a Federally funded minor use pesticide registration program,
      with both field subcontracting and residue laboratory phases.
  c. several regional research grants dealing with pesticide
      residue fate and metabolism, groundwater migration, applicator
      exposure, rinse water treatment and disposal, and museum
      worker exposure to preservatives.
  d. multiresidue methods development and analytical contracts with
      industry, other departments at Cornell, groundwater programs
      at several northeastern universities, and emergency analyses
      for local private operations having suspected leaks, spills or
      misapplications.
  e. methods development, analysis and basic research on con-
      tamination in wines and spirits -- investigations sponsored by
      both private industries, and producer cooperatives.
  f. numerous cooperative studies with other Cornell faculty,
      ranging in duration from several weeks to several years.

No one program, regulatory or otherwise, has exceeded 30% of total budget. And the diversification in training and experience our staff has received allows them to move between several projects as funds and workload dictate. Consequently, no particular area of endeavor exerts a disproportional influence on priorities, or procedures.

While sample documentation and tracking protocols, as well as QA/QC procedures, have been in place and in effect for many years, their application is tailored to the needs and resources of each individual project, with some exploratory or field-feedback studies requiring much less verification than other regulatory or complaint series. Flexibility in experimental design is imperative if a facet of an investigation is to be completed in accordance with its significance and allocated resources. However, since several of our programs are potentially answerable to GLP´s, it was deemed important that the laboratories current level of compliance be established so as to accurately measure the full fiscal and scientific impact to be felt if we should attempt to institute a formal GLP structure.

To this end, an informal audit was conducted, by EPA personnel, of the laboratories in general, and of some specific studies from which residue data had gone to EPA and been used in regulatory decision making. Some of that report will be cited verbatum in the next section. However, since subsequent interpretations, alternatives and opinions voiced in this presentation are those of the author or his academic colleagues, the EPA personnel declined to be formally associated with this preparation and presentation of the case history. We acknowledge their efforts and input to this paper, and we recognize that while they may not agree with some of its content; they will have their views expressed in other units of this series.

THE AUDIT

Deletions from the original report herein cited are to provide autonomy to personnel and projects and are not made to enhance or diminish the position of either party with regard to the central question.

"On February 10 and 11 (1987, we) met in Geneva, New York and informally audited the Good Laboratory Practices (GLP´s) and performed a Quality Assurance (QA) audit of ... projects chosen at random. It was intended for the laboratory personnel to understand that the audit process would be helpful to their organization and also helpful to the ... program. The entire staff of the Geneva labs participated in the process in a very cooperative and hospitable manner.

The audit was divided into two parts: (1) General QA Practices and (2) Data Audit. A questionnaire as used by EPA´s Office of Pesticide Programs, Quality Assurance Office, for Internal Audits, was employed. A copy of the completed questionnaire is(not) attached. From discussions, based on the questionnaire and observations made during inspection of the facilities and from a data audit, a subjective summary regarding

general laboratory practices, and integrity of specific data follows.

I. General Quality Assurance Practices
   A. Organization of the Analytical Laboratories is well defined
      by area of responsibility into two sections:
      1. Pesticide-Toxic Chemicals Section; and
      2. Feed and Fertilizer Section.
      However, the laboratories share common facilities such as
      shipping and receiving, purchasing, storage space, etc.,
      with the entire Food Science and Technology Department.
   B. Quality Assurance Program is the responsibility of a QA
      Officer who in turn is chairperson of a QA Committee. The
      elements of the QA Program are presented in the "Laboratory
      Quality Assurance Manual." August 1986.
   C. Personnel are knowledgeable and well-trained to do the
      assigned tasks; equipment and facilities are adequate.
      Turnover of personnel is very low as compared to similar
      laboratories; therefore, experience of even junior person-
      nel is far above average.
   D. Attitude of Management and staff to GLP's and QA is very
      good. Everyone recognizes the need and appears to be
      interested in implementing necessary practices to assure
      integrity of data.
   E. The laboratory audit and data audit were concerned with
      only the Pesticide-Toxic Chemicals Section of the
      laboratory. General comments on these areas of respon-
      sibilities follow:
      1. Studies or projects do not have a plan for QA which was
         reviewed/administered by the QA Officer. (Some dis-
         cussions indicated that the QA Officer has not been too
         involved with pesticide analyses.)
      2. Methods of sample handling are well documented and
         assure a good trail-of-evidence from time of sample
         receipt at the laboratory to time of the report audit.
         However, information received from the field for in-
         dividual samples is often incomplete.
      3. Maintenance records of reference standards, stock stan-
         dard solutions, and working standards are not complete,
         and a method of handling these standards is not properly
         documented. Storage of stock and working standards,
         labeling of all standards, and a plan for disposal are
         not well defined.
      4. Instrument SOP's are not documented; however, instrument
         logs are well maintained with the possible exception of
         GLC detectors.
      5. Solvents and reagents are well maintained, but they do
         not always carry labeling as to time of receipt and
         storage data. Condition of distilled water is ques-
         tionable.
      6. Use of bound laboratory notebooks occurs in the Feed and
         Fertilizer Section but not in the Pesticides Section.
         Data trail proceeding from final report backwards to
         original chromatograms (GLC and HPLC) or spectrometer

records was adequate enough to debate whether there is a need for bound records in the Pesticides work. However, there is no substitute for a diary record to assure that every problem or success is documented.

7. Freezer storage of record samples and working samples is good, but could possibly be improved if separate facilities for the Analytical Labs were available. Documentation of storage temperatures is needed. Some provisions should be made ... to notify the lab when record samples can be discarded.

F. Recommendations on General QA Practices

1. Internal audits should be periodically scheduled by the QA Committee and QA Officer, probably at intervals of three months, using all qualified personnel in audit teams. This procedure will increase awareness, promote training, and share responsibility for GLP's with each individual.

2. The QA Officer should serve as a monitor on the quality of data (all data) that are reported from the laboratory. The QA Officer should inspire fellow workers to work diligently for the integrity of all data.

3. The Analytical Laboratory staff should become involved in implementing the entire contents of the "Laboratory Quality Assurance Manual," as written, or change those criteria that are impossible or of no value.

4. All laboratory personnel should be encouraged to visit other laboratories, to attend scientific meetings, and to talk with other analysts with particular regard to GLP's. Such interchange will allow them to learn that the Geneva labs are probably above average in the chemical community.

II. Data Audit

Two ... projects were chosen with the following criteria in mind:

1. Different method of analysis − HPLC and GLC;
2. Different individual analysts; and
3. Data had been received by EPA/RCB.

With the above criteria in mind, we audited:

1. ... − Benomyl on Chinese Cabbage ... and
2. ... − Fenvalerate on Beets ... .

The analyses for benomyl residues in Chinese Cabbage were performed by R. A. Marafioti in 1982 using the HPLC procedure as published in the J. of Chromatography, 317(1984) 527–531 (Spittler, T. D.; Marafioti, R. A.; Lahr, L. M.). Successive determinations of MBC and 2−AB are achieved, with MBC residues calculated as benomyl.

From final report backwards to the original report and record of sample, the trail was found to be easy to follow. Copies of chromatograms appearing in the final report were matched with original chromatograms. Calculations of data were determined to be correct. However, an outstanding problem of transposition of data

occurred from the original calculations to the reported concentrations of residues, in check samples, only. All check samples were
reported as ppm of 2-AB equal to total residue with the concentration of apparent benomyl deleted. The data as reported and as
calculated are:

Sample Number	Application Rate	(As Reported) Benomyl (ppm)	2-AB (ppm)	(As Calculated) Benomyl (ppm)	Total Benomyl (ppm)
290	0.0 lb/ai/A	0.07	0.07	2.59	2.66
291	0.0 lb/ai/A	<0.02	<0.02	3.33	3.33
292	0.0 lb/ai/A	<0.02	<0.02	0.07	0.07
293	0.0 lb/ai/a	<0.02	<0.02	0.06	0.06

Since the analyst was aware that the check samples were
contaminated, he used Sample No. 293 for his spiked recovery and
the 0.06 ppm as background residue in the check sample. Since the
residue concentrations in treated samples ranged from 3.2 ppm to
8.1 ppm, integrity of the data or its usefulness in the report, as
reviewed by EPA, was not compromised. A careful look at the
chromatograms as copied for the final report and the original
chromatograms as calculated by the analyst indicated that the
person preparing the report did not interpret the (tabulated) HPLC
data correctly. Thus, anyone else who might look at the numerical
values versus chromatographed data could be a victim of misin-
terpretation.

It is apparent that a person such as a QA Officer should
monitor the data that have been prepared for the final report.

Since the trail-of-evidence was able to be reconstructed so
well and the analyst performed his tasks correctly, the value of
the data was not lost. However, this circumstance serves as a good
model for supporting the necessity of GLP's (which were mostly in
place) and QA procedures (which were not followed).

..., Fenvalerate on Beets, was analyzed in 1982 by G. Helfman
using the GLC procedure for the parent compound, only. The
reported data, calculated data, copies and original chromatograms,
analytical methodology, record samples, and field data description
were easily found, followed, and determined to be acceptable.

In conclusion, the Data Audit was considered to be successful. One study was found to be acceptable and another study showed
the need for improved QA practices. Fortunately, GLP's were of
sufficient quality that no loss or compromise of data was experienced."

IMPLICATIONS

The impression, at first blush, is that with minor changes, the
laboratory could fully meet GLP's, in fact, it is almost in compliance, already. Thus, it would be relatively simple, and to the
advantage of the laboratory and its programs, to institute the
necessary measures. Wrong. There are several subtle aspects that
prevent our laboratory from accepting GLP standards, and there are

additional reasons why strict adoption at the university level is
not only unnecessary, but undesirable.
   Few, if any, university regulatory programs are funded at a
level sufficient to maintain GLP's. Most depend on state allocations or contributions as overhead to maintain what is usually an
operation beneficial to the state in which the university is located. Thus not only is the extra manpower not budgeted, there are
no provisions for separate sample preparation and storage
facilities, separate freezers, locked and inaccessible record
archives and equipment. In fact, many of the dictates for maintaining locked space are in direct conflict with department and
university policy. Blocks of valuable space will not be reserved
for the exclusive use of these (or any) programs. Even storage
space for data and documentation is difficult to justify or obtain
in light of the productive uses that compete for room.
   The concept that every sample in a laboratory have a unique
history, identification, location and destination is basic to GLP.
That there might also be unrelated samples within the facility that
do not have these extensive pedigrees could constitute a violation.
Yet, many items do not need or warrant GLP tracking; are they to be
excluded from a laboratory operating under GLP's? Or, conversely,
will a laboratory having such samples on premises be judged as no
longer being in compliance.
   Just as good government rules only with the consent of the
governed, the benefits of GLP's exist only if the tenets are accepted by the affected scientists. But, for the university
community to accept them would be not only redundant but counter
productive. We are keenly aware that many of the national efforts
in which we participate, efforts sited throughout the land-grant
college system and other universities, rely on the research at our
academic institutions for data and information essential to decision making processes. Many times these are the primary source of
information, particularly when preliminary assessment or emergency
response situations occur. Most of the research producing these
data bases is not, and never will be, conducted under the GLP
dogmas. These studies are designed and executed to pass the
scrutiny of the investigator's peers, that is, to be accepted as
valid and reproducible, meeting the standards of the discipline.
Implicit to the standards of a discipline are the assumptions that
a study be conducted with competence and integrity and be a credit
to the research group. Of course, intentional deceit is possible,
reports, data and conclusions can be embellished or falsified; but,
this can also be done under GLP conditions, with proper attention
to detail.
   A statutory set of GLP regulations would be necessary and
reasonable if intended for a totally naive laboratory having no
other code of conduct, general guidelines or direction. Most of
this is already in place at a university. Various levels of technical and departmental supervision and oversight exist, and a
group's output is routinely submitted for peer review -- submitted
voluntarily so as to establish and maintain the investigator's
reputation and credentials. There is no percentage in, or incen-
tive for, an academic group to submit questionable or falsified
data for regulatory consideration. Accordingly, the reputation of

the principal investigator, his department and university should
be, and must be, an acceptable substitute for the field and
laboratory GLP provisions. EPA would have the option of refusing
to accept data from a particular source, but, they would also have
to have cause, and be prepared to document and defend their decisions. Ideally, this type of assessment should be made by an
independent panel if and when a data source becomes suspect. It is
doubtful that EPA can establish and maintain a program of data
verification better than what already exists as a result of
departmental oversight, peer review for publication, and professional integrity. What is more, given their record of
reorganization and turnover they would probably do much worse, and
waste a lot of other groups resources in the process.

   This brings us to the question of economic impact: it has
been estimated -- DRAFT form, of course -- that the increased cost
of investigations under full GLP´s will be 20%. We can agree with
this if we assume that all facilities modifications and additions
have been made, and that the initialization costs have already been
covered. However, this estimate is deceptive. For instance:
field GLP´s are a necessity if laboratory GLP´s are to be meaning-
ful. Very few university laboratory programs do not rely on
university affiliated field cooperators or contracts for one or
more aspects of a given investigation. Usually there are numerous
field operators for each laboratory: the cost of GLP compliance
has now been multiplied many fold, particularly when one realizes
that frequently sample production is undertaken voluntarily, or in
conjunction with related activities. Few of these field
cooperators upon which our programs have been imposing could, or
would, be able to absorb the additional effort or facilities necessary -- a costly alternative would be to go to private contracting.

   There is also no grandfather clause for studies completed or
in progress or in the extant literature. Unfortunately, no one
considered the costs of repeating everything that did not anticipate their good ideas. An additional unforeseen consequence is
that repeated studies will have to compete directly with new initiatives for program space and resources. Delays will result in
economic losses to producers and manufacturers, and in increased
costs to consumers. Apparently, none of these points are considered important enough to make it into a cost/benefit statement.

   Also, if the insistence is made that only GLP approved
studies are acceptable in support of a chemical or its use pattern,
will these same standards of scrutiny be applied to studies containing findings that reflect negatively on a chemical? Will any
reports of toxic or carcinogenic effects be automatically ignored
by EPA if they haven´t been conducted under verifiable GLP conditions? What would be the whistle blowers reaction to that? Does
anyone really believe EPA will apply the same standards to negative
reports that they insist upon for evidence submitted as support?

CONCLUSIONS

We concede that EPA-directed research should be subject to their
mandates, assuming they are also paying the costs for meeting those
mandates. We also recognize that private laboratories must meet

their expenses and generate a return to their investors, to do this they add on the cost of GLP compliance and facilities to study estimates. The Chemical Industry, too, knows that if it must go along, compliance is added overhead and can be passed on to their clients and customers. Even other agencies can absorb the expense of new rules and procedures by reorganizing and cutting back on productivity. Unfortunately, university funding has no escalator clause for new rules, no mechanism for covering elevated costs. Our response has been to increase budget requests; these requests have been for the most part ignored. Our sponsoring agencies function under conditions that do not allow funding at anywhere near the levels routine to EPA and apparently necessary to meet their dictates. Until this discrepancy is resolved, either by statute or understanding, many of our programs are at an impasse. For the universities, a point of diminishing returns is being reached. If GLP compliance is mandated, and if such mandates greatly increase overall program costs, and if such costs cannot be covered by increased operating allocations, the output of the programs dwindle. However, a program diminished by the expense of superfluous requirements is no longer worth the attention of the university or the facility it occupies.

The consequence is not that regulatory agencies will lose all university participation in their programs; they will only lose the interest of the good ones. As better organizations go into more productive areas, mediocre and less qualified groups will come in to take their place -- groups that would accept almost any program. With scientists as with bureaucrats, you can never set standards so low that a population can't be found to fit them.

The alternatives from the academic perspective are obvious. We will decline to recognize GLP mandates for regulatory studies and research conducted under university tutelage until or unless the agencies involved recognize the necessity and responsibly for maintaining support at levels sufficient to meet their own require- ments. In addition the value of existing data bases and literature must be upheld.

RECEIVED January 29, 1988

Chapter 17

Good Laboratory Practice Standards in a University Setting

Problems and Solutions

Willis B. Wheeler [1] and Neal P. Thompson [2]

[1]Pesticide Research Laboratory, Institute of Food and Agricultural Sciences, University of Florida, Gainesville, FL 32611
[2]Office of the Dean for Research, Institute of Food and Agricultural Sciences, University of Florida, Gainesville, FL 32611

The University is fundamentally an institution of learning, although research and service by its faculty are certainly of major significance. In addition, the land-grant academic institutions have a mission to help their state realize maximum potential for agricultural development and to contribute to the solution of social, economic, environmental, and cultural problems of concern to the citizens of that state. These missions are carried out through the three closely related functions of resident instruction, research, and extension.

Good science exists in research in academic institutions apart from the soon to be established Environmental Protection Agency's (EPA) Good Laboratory Practice (GLP) guidelines. However, this is not to say that GLP's are not advisable. Good science should be able to stand up to review as having been performed using appropriate and adequate laboratory practices. Scientists have had their work routinely scrutinized by their peers for its quality and will not resent careful analysis by others. For example, only a portion of the work produced by the scientific community is acceptable for publication in its various journals. The rate of acceptance in journals varies but it is apparent that the peer review system attempts to serve as a quality control mechanism in the scientific community.

As a practical matter, however, much of the data to support various pesticide clearances does not come from the peer reviewed scientific literature. The original methods undoubtedly have been subjected to such review but the laboratory practices used to produce the data are often unpublished modifications. GLP guidelines, therefore, are intended to assure that the science used as the basis for regulatory decisions is reproducible. Scientists do not need GLP's to achieve accurate results but GLP's assure the public and their representatives, the regulatory agencies, that adequate practices are in place. The general public has every right to know that the products of scientific enterprise are of high quality. The regulatory agencies perform this service. It is important for scientists and the public to be reminded that regulatory agencies exist in the public interest and not as an entity for their own fulfillment.

0097–6156/88/0369–0126$06.00/0
© 1988 American Chemical Society

For a number of reasons, GLP's are to be implemented and will apply to all groups that develop data in support of a marketing permit for a pesticide product. Many institutions of higher learning, will come under these proposed regulations as they are currently written. There are at least three possible approaches to addressing the situation: attempt to satisfy all the requirements; attempt to satisfy those requirements that can easily be implemented at reasonable cost and negotiate with the EPA on those aspects that are very difficult to implement; and choose not to comply. This paper will approach this situation from the viewpoint of finding approaches to satisfy the major components of the GLP Standards.

There are several areas in the GLP guidelines which are more difficult to implement in an academic setting than in other settings. These are: requirements for a separate Quality Assurance Unit (QAU) and the responsibilities thereof; establishing Standard Operating Procedures (SOP's) and conformance to them as written; and the significant added costs of doing business under the GLP's.

The QAU is responsible for monitoring each study to assure that the facilities, equipment, personnel, methods, practices, records and controls conform to GLP regulations. "For any given study, the QAU shall be entirely separate from and independent of the personnel engaged in the direct conduct of that study." (Quoted from the EPA's proposed Good Laboratory Practices Standards.) The QAU must "inspect each study at intervals adequate to ensure the integrity of the study...," and "determine that no deviations from approved protocols or standard operating procedures were made without proper authorization or documentation." (Quoted from the EPA's proposed Good Laboratory Practices Standards.)

It is unfortunate that the proposed regulations are expressed in such a manner as to suggest a lack of confidence in laboratory and field work. Trust and integrity are a major tenet of the academic community. In the vast majority of cases, academics provide solid, scientifically valid information which undergoes and survives peer review in the absence of QAU's and SOP's.

There are many possible approaches to establishing a QAU; the authors will suggest a few which may be feasible approaches at our institution. At the University of Florida, the Institute of Food and Agricultural Sciences (IFAS) comprises the research, teaching and extension components of this land-grant institution. There is a Vice President, a Dean for Research, a Dean for Instruction and a Dean for Extension; in addition there are some 40 department chairmen and unit heads (heads of IFAS units located throughout the state of Florida). With this kind of structure, one QAU could assist all units that would require such service. In fact, considerable research goes on within Florida that ultimately bears on the registration of pesticides. The size and structure of the QAU would depend upon the number of projects that have to be monitored. The director of that QAU could be drawn from the faculty or from outside sources. The director might report to the Dean for Research and as a result could function independent of pressures that might be imposed on him/her if he were a department faculty member or a lower level employee. With adequate support personnel and budget, the QAU director would be able to certify that GLP's were being adhered to throughout the IFAS statewide system.

A serious impediment to establishing such a QAU is cost. Support might come from grant indirect costs and/or if the impacted federal agencies would agree, as direct costs of doing business. The authors would even suggest that in cases where no indirect costs are allowed by granting agencies, that the EPA consider some financial assistance in establishing QAU's.

If the QAU need not service such a large operating unit, it might be housed in a single department and staffed by an individual who would report directly to the department chairman. This again would remove that individual from potential pressures that could be brought to bear as a result of his/her duties. Having to service a departmental program might, however, have very high "costs." If the QAU director were a member of the faculty, the time that this job would require could adversely affect his ability to perform the functions upon which he is evaluated for promotion, tenure and salary increases. Further, such responsibilities would consume time needed to further a career and might be considered by many as a highly undesirable duty. This might then suggest, that a senior faculty member, whose research program had slowed somewhat, might be a candidate for such a position. The impact on the career of a senior faculty member who might be approaching retirement would be minimal. In the case of such a small scale QAU, the costs involved could also be more easily absorbed.

Such a departmental system could be established and implemented as needed. Thus each department or unit could operate a QAU perhaps with some support from the administration or from those agencies that have established the requirement to have such units. A negative aspect of such a system could be the variability of the QAU's. With a large number of individuals involved there would likely be great variation in the quality of the QAU. The authors are not trying to be facetious, but in such a situation, one would almost need a super-QAU to oversee the operation of the smaller ones. As a result, once an organization required more than two, three or perhaps four such units, it should probably establish one larger QAU for the entire organization.

Selection of QAU directors is also very critical. Some regulators function in a rigid manner, seeing only right or wrong. The vast majority, fortunately, are willing to work with those whom they regulate and provide assistance to achieve common goals. As long as both parties are willing to work to achieve the same objectives, with some patience and understanding, then usually those objectives can be achieved. Having a QAU director who understands academics and the academic ways, will be important in establishing QAUs and to their ability to function effectively.

The second area mandated by GLP's is establishing SOP's and the requirement to adhere or revise them. One of the functions of the QAU is to assure that the SOP's are in place and being followed. The negative aspects of the concept of SOP's are the absolute standardization of everything that is done in any setting. A university is one of the few places where creativity does and must exist. The two concepts are diametrically opposed and as a result, the SOP's will be difficult to get established and become accepted.

Another negative aspect is the time, energy and resources that will be required just to establish and implement the SOP's.

Approaches to satisfying the need for having SOP's are several. One may start from scratch and write out procedures for everything from writing "date received" on reagents to interpreting mass spectral analyses. A more feasible approach would be to adopt a set of SOP's created by the EPA that would satisfy the needs of the Agency. Each laboratory would have to adapt such SOP's to its own situation and tailor them to their own needs. In addition, modifications on a laboratory by laboratory case would have to be incorporated to achieve the appropriate goals. On the whole, however, a model set of SOP's from the EPA would greatly assist university laboratories in establishing their own and reduce the drudgery of the task.

The third significant issue is the financial commitment involved in adhering to GLP's. These have been alluded to above. The economic impact of these proposed regulations, based upon the EPA projections is as follows: "...testing costs will increase by about 20 percent." (Quoted from the EPA's proposed Good Laboratory Practices Standards.) The private sector can and will pass these added costs on to the consumers. The public universities, however, cannot pass the costs on to their consumers and will have to pay the added expenses from some other funding source. One possible source would be indirect cost increases to the agencies funding research that would fall under these regulations. Some projects that contribute data directly to the EPA, specifically to support the registration of pesticides, do not collect any indirect costs from their funding sources. If an institution is currently paying the administrative costs of a grant without the benefit of indirect costs to help defray those expenses, it is unlikely that institution would be in a position to provide an additional 20 percent of a grant's operating budget to allow the program to remain at its level of productivity.

The authors suggest, therefore, that the EPA consider funding a portion of the expenses related to GLP's. There are two significant costs involved: The first is the initial set-up stage where the SOP's are established and implemented; and the second is the actual performance under the GLP regulations. The 20 percent increased costs probably only apply to the actual performance under GLP regulations. The initial cost of establishing SOP's and preparing to adhere to GLP's will be significant. One estimate of the time required to accomplish this stage is six months. If the EPA would entertain the possibility of assisting those institutions that have no other source of funding, perhaps through a grant and/or the preparation of a set of malleable SOP's, this could alleviate the bothersome and costly aspects of this exercise. The EPA assistance in getting set up, according to the needs and desires of the Agency, will certainly be worth any costs incurred.

The other costs (i.e., that 20 percent day-to-day operating cost) will have to be paid in lost productivity. Funding agencies often have fixed dollar amounts for research projects. Whether costs of adhering to GLP's are charged as indirect or direct costs doesn't matter. The total budget determines the amount of effort that can go into a project. Adherence to GLP's will become a part

of the mounting list of fixed costs of doing business; the flexible
components of grant supported projects will be reduced.

It is unfortunate that public universities who are performing
a public service will be required to comply with GLP's. It is also
unfortunate that proposed GLP's are written in a way that may ali-
enate the faculty that must try and conform. Some institutions may
refuse to comply; as a result, those who have made and might continue to make important and significant contributions to the those
agencies that are imposing the regulations, will be lost. The
authors suspect that the EPA will lose more than it gains.

Those of us who dislike the impending GLP's but can understand
the EPA's need for such regulations will make a good faith effort
to implement and follow those regulations. The EPA can make our
job much easier and less distasteful by providing financial and
other assistance and by exhibiting patience and understanding.

Academic institutions are facing difficult fiscal times as are
many other areas of our society. The research universities are
often being asked to do more with less. Even in these days of
reduced inflation, research support is a precious commodity.
Equipment costs rise, personnel costs rise, and the general cost of
doing business increases with the net result that less research is
accomplished. Unreasonable increases in paperwork reduce actual
productivity.

At a time when increased productivity must be a goal in light
of increased competition for funds, increasing regulation appears
to be counterproductive. This is particularly true in the public
sector where much of the supporting data is for "minor" use, that
is those uses that are beneficial to the public but bring little or
no economic benefit to the industry. We dwell in a society which
encourages incentives leading to economic success. The public sector represented by the public university plays a beneficial role to
society which cannot be linked to profit.

The intention of the GLP's is good, but must be tempered with
the practical reality of academic research and education. Several
scenarios could be advanced such as tampering, theft and falsification of data which would support the need for more regulation.
Falsification of data has occurred as in the much publicized and
extremely damaging IBT case. The university climate, however, must
be such that innovation and creativity are fostered in a collegial
atmosphere; increasing limitations by way of regulations are detri-
mental to this atmosphere.

RECEIVED January 29, 1988

Chapter 18

Quality Assurance for Ecotoxicology Studies

John A. McCann

U.S. Environmental Protection Agency, EN-342,
Washington, DC 20460

Ecotoxicology involves the study of the
effect of toxic substances on mammals and
birds in their environment, and aquatic
organisms in fresh and salt water. This
paper will discuss the quality assurance
aspects of studies involving direct
application of potentially toxic materials
to test organisms. It will stress the need
for chemists to become more involved in
ecotoxicology testing by assisting
biologists in documenting the identity of
test substances, the exposure levels and
the stability of the test material in
water, air and/or food, and in measuring
residue levels in the test organisms and
their surrounding environment.
    Good Laboratory Practice and Quality
Assurance procedures required for acute and
chronic health effect studies can be used
for acute and chronic mammalian, aquatic
and avian ecotoxicology studies.

Publication of the Environmental Protection Agency's Good Laboratory
Practice (GLP) regulations governing the conduct of studies submitted to the Agency in support of The Federal Insecticide, Fungicide and Rodenticide Act (FIFRA) and the Toxic Substances Control
Act (TSCA) has caught the attention of the regulated public. An
important part of these regulations assures that the laboratory and
the studies meet GLP requirements by describing the activities of
the Quality Assurance Officer and Quality Assurance Unit at the
laboratory.
    This paper discusses the quality assurance procedures for eco-
toxicology laboratories. It will attempt to concentrate on those
areas that are of particular concern to those individuals determining the toxicity of chemicals to fish and wildlife (ecotoxicology
testing).

A minimum amount of time will be spent on those areas that are common to all studies and which will have been covered more extensively by other authors.

The requirements for ecotoxicology studies vary little from the requirements for health effect studies. Laboratories that have adequate facilities, equipment, staff, and procedures for health effect studies, with adequate training of their staff in specific areas of ecotoxicology, could conduct these tests.

For example, the requirements for tracking the receipt and use of the test substances and test animals are the same. The need for the calibration of the test equipment, storage, and archives are the same.

Extra effort should be expended, however, to determine that adequate good quality water is available at aquatic testing facilities by more frequent analyses of incoming water. Special arrangements should also be made to handle large volumes of waste water such as pretreatment of the water using charcoal filters before discharging the water from the facility.

Because the aquatic studies require a more precise evaluation of the quality of the incoming water and the use of specialized equipment, I will spend more time discussing this area.

Aquatic Environments

Fish and aquatic invertebrates are exposed to toxic substances in the laboratory by one of three types of systems: static, flow-through and renewal, as described below.

Static

In static tests the test material is mixed with the water. Then the aquatic organisms are placed in the test solution and remain there for the duration of the test or until they die. The tests normally last for two days for small invertebrates and four days for fish, amphibians and larger invertebrates. The experiments are generally considered "quick and dirty" tests that give a reasonable estimate of the toxicity of the test substance with a minimum of effort. In the past, they did not require analysis of test material concentrations unless solubility limits were being exceeded.

Flow-through

In flow-through tests, the treated water is continually replaced either by a constant flow or by additions of small volumes of treated water at 1-10 minute intervals. Various delivery systems have been designed to either supply measured amounts of newly mixed test concentrations or to add premixed solutions to the test chambers. Depending on the physical characteristics of the chemicals involved and the reliability or accuracy of the delivery systems, there is ample opportunity with this method for something to go wrong in the delivery of the test chemical. However, because the water is constantly replaced and the organisms can be fed, it is possible in tests conducted in flow-through systems, to expose test

organisms to the test solutions for several years. Because of the
importance put on these tests by regulatory agencies and the potential for fluctuation of test concentrations, it is generally
considered necessary to measure daily and document the concentration
of the test substance in the water in at least the high, medium and
low concentrations. If control and solvent controls are also
included in the sampling scheme, the number of analyses could involve
over 1500 samples per year per study.

Flow-through systems can be used for acute studies but generally
they are used for uptake and depuration studies or complete life-
cycle studies.

Renewal

The renewal system is a combination of the above systems. The test
organisms are removed from the old test solutions and are placed in
new, freshly prepared solutions of the same concentrations three
times each week. A representative sample of the old and new solutions at high, medium and low concentrations are analyzed each time
the transfers occur. This system is frequently used for the daphnia
life-cycle tests. It is also used when the test chemical has a
short half-life in water or if the test animals must be fed during
the test.

Length of Tests

Another area I want to discuss is the length of ecotoxicology
studies. This generally is dependant on the type of test.

Acute Studies

Acute aquatic tests normally last 2 to 4 days depending on the test
organisms. Chronic tests for invertebrates, like daphnia, last for
21 to 28 days and may involve several generations of offspring. We
are always looking for test organisms that will reach maturity faster
so we can evaluate the effect of the test substance on multiple
generations in a shorter period of time.

Chronic and Subchronic Studies

Full life-cycle or chronic fish studies may take two years or more.
Consequently, biologists and regulators are requesting more subchronic studies where the test organisms are subjected to toxic
concentrations during their early life stages (egg to fry) for
generally a 30-day exposure period. If there are multiple test
concentrations and the turnover rate for the water is rapid enough,
a large number of analyses and extensive record keeping may be
required.

For chemical analyses to be useful to the biologist, analyses
must be timely. A significant delay in supplying a water or food
analysis to a biologist could invalidate a study if the deviation
from nominal concentration is shown to be too drastic or for too
long a duration.

Most of the ecotoxicology tests involving birds and/or small

mammals have the same requirements as the rat and mouse tests
involved in routine toxicity testing and require the same analysis
of the food and tissue.

Special Considerations

Now that we have discussed the terminology and the types of tests
involved, I want to emphasize some areas that should be well documented in an ecotoxicology study.

I am not going to go into the more obvious quality assurance
program requirements that are applicable to all laboratories. These
general requirements include the provision for a Quality Assurance
Unit, the availability of a qualified staff, the presence of facilities and equipment adequate to permit the type and number of
studies being performed, a master schedule of ongoing and completed
studies, storage areas for active and used samples, and archives for
the retention of reports and raw data.

With the exception of the special handling required for the
test water, wild birds and fish, facilities that conduct routine
mammalian toxicity studies should be adequate to conduct ecotoxicology testing.

Since we are talking about testing that is frequently conducted
on wild species that have not been routinely tested in the laboratory
it is important to stress that Quality Assurance personnel be aware
of special requirements, such as temperature control, light control,
cage or tank size, water quality, etc. This information should be
available in the protocol and the SOP.

Again, the environmental and testing conditions for many wild
mammalian and avian species are compatible with the domestic
mammalian studies that have been done in some laboratories for years.
Since some of the ecotoxicology studies concentrate on the nonlethal
effects of the test substance on the test organisms, it is important
that the test conditions and evaluation criteria be accurately
described and the staff be very aware of sublethal effects of the
toxicant on the test species.

For aquatic studies, it is important to frequently document the
quality of the incoming water, the quality of the water in which the
test organisms are held and acclimated, and the quality of the water
in which the animals are tested. The temperature, alkalinity, pH,
hardness, salinity, etc., may be well within the criteria that is
acceptable for mammals and birds but not aquatic animals. For
example, the abrupt changes that could occur when fish and/or
invertebrates are transferred from one water quality to another,
could either stress or kill the animals. Low chlorine or mineral
levels in water might be acceptable for birds and mammals but may
be deadly to fish or invertebrates. Raw water coming into an
aquatic laboratory should be analyzed for water quality and chemical
residues at least quarterly, until a data base has been established
that demonstrates that the water quality falls within acceptable
parameters and contains no significant contaminants. Semi-annual
checks should then be used to confirm the continued acceptability of
the water. The laboratory should routinely document the quality of
water being used to hold, acclimate and test the aquatic organisms
to assure that the water is acceptable for each test organism.

Types of Ecotoxicology Studies

There are approximately 15 to 20 types of ecotoxicology studies
that are normally requested by EPA. I am not going to describe them
all, but I do want to familiarize you with some of them (Table 1).

Table 1. Representative List of Environmental
Toxicology Studies

1. Daphnid acute toxicity test
2. Daphnid chronic toxicity test (life-cycle)
3. Mysid shrimp acute toxicity test
4. Mysid shrimp chronic toxicity test
5. Oyster acute toxicity test
6. Oyster bioconcentration test
7. Oyster shell growth test
8. Penaeid shrimp toxicity test
9. Acute fish test (cold and warm water) fresh and
      salt
10. Fish bioconcentration tests (fresh water)
11. Fish early life stage test (fresh and salt water)
12. Avian dietary test (cold and warm water) fresh and
      salt
13. Avian reproduction test (mallard, bobwhite quail)
14. Wild rodent 5 day feeding test
15. Wild mammal tests (skunks, wolves, foxes, rodents)
16. Special avian and mammalian tests
17. Algal acute toxicity tests
18. Seed germination/root elongation toxicity test
19. Plant uptake and translocation test
20. Small pen/field studies

Biologists frequently use daphnia in freshwater short and long
term tests, while shrimp and oysters are used to evaluate the
potential toxicity of chemicals to saltwater invertebrates.
Freshwater fish tests are generally conducted on bluegill, a
warm water fish, and rainbow trout, a cold water fish. Catfish, fathead minnows and sometimes carp are also used depending on the
expected route of exposure. Sheepshead minnow is the commonly used
saltwater fish.
Chronic studies are conducted on the fast-maturing fathead
minnow and sheepshead minnow. Subchronic studies are conducted on
those species relatively easily raised in the laboratory; e.g.,
fathead minnow, trout and sheepshead minnow.
The quality of the incoming water is of particular concern to
individuals conducting aquatic studies. Potable water is generally
recognized as poor quality water for fish and invertebrates. A
good fresh water is one in which daphnia will live and satis-
factorily reproduce.
Avian LD50, dietary LC50, and reproduction studies are normally
conducted on young or adult mallard ducks or bobwhite quail,
depending on the requirements of the test.
There has not been the same demand for ecotoxicology testing of
mammals because the Agency has routinely extended the results of

routine (health effect) toxicology tests to wild mammals. A five-day feeding test and other special tests have been designed to give the environmental toxicologist a better idea of field exposures of mammals and birds to toxic materials.

Botanical tests would be evaluated using similar criteria as those used on other ecotoxicity tests.

Analytical Support

There are several areas in aquatic toxicology where chemistry support would be very helpful to the biologist conducting the study. In other areas, chemistry support is essential, e.g., analyses of test concentrations and residues in test organisms, feed and water. If chemistry support is not available to the biologist, the biologist should have considerable expertise in these chemistry oriented areas. I hope to encourage ecotoxicology testing facilities to supply the required chemistry support to their biologists.

Test Material

I am going to start the discussion by tracking a test chemical from its arrival in the laboratory to its final archiving and storage. However, as you will see, the required analyses are not limited to analysis of the test substance.

When the test material arrives at a facility, there should be a fact sheet with it that leaves no doubt as to the identity of the test material, lot or batch number, percent active ingredient, storage conditions, etc. The weighing of the incoming sample can be determined by a technician. The weight of the sample will serve as the starting point for the use log for the chemical. In several laboratories, I have seen chemists go several steps further. They sometimes confirm the identity and purity of the sample. They sometimes indicate the solvents of preference and the solubility of the material in the various solvents and water. This aids the biologist in selecting test concentrations and appropriate solvents and does not create unnecessary problems with chemical determinations later because the wrong solvents were used. Obviously, all the calculations, measurements, etc., must be adequately recorded to satisfy GLP requirements.

The adequately labeled containers of test material must be stored in appropriately labeled areas under conditions that will have no adverse affect on the chemical's stability, composition, etc.

Water Analysis

If surface scums or precipitates are observed in an acute study or the protocol requires it, the test concentrations need to be measured and documented. This requires taking appropriate samples. The investigator must have considerable expertise in taking and analyzing the samples if the concentrations are low or if sophisticated analytical techniques and/or equipment are needed. In a renewal-study or a chronic study requiring a flow-through system, it is important to take and measure test concentrations on a daily

basis so that timely corrections can be made in the delivery system
to maintain test concentrations. If the samples are not analyzed in
a timely manner, the biologist may not be able to maintain the test
concentrations. If the deviation from the intended concentration
occurs for too long or is too severe it could invalidate the study.
Many of these chemical analyses may be too difficult for the average
biologist.

Test Solutions

Some chemistry units support the biologists by preparing the test
solutions. Whoever prepares the aliquots of test material and/or
makes the stock solutions should adequately document the removal of
the appropriate amounts of test material from the bulk container.
The balance readings used to weigh out the samples should become a
permanent record and should be retained as original raw data in the
archives.
      The archives should be adequate to maintain the identity and
integrity of the sample. In some cases, this could require freezing
the sample.

Residue Analysis

Chemistry support is frequently needed to measure the residues of
possible toxic substances such as pesticides in each lot of test
organisms before they are used in toxicity tests. The analysis
should be extensive enough that the data will document any residues
that might interfere with the usefulness of the test.
      Inspectors and quality assurance personnel should routinely ask
to see the chemical analyses or residue analyses performed on each
lot of test organisms used in bioassays that were reported to the
Agency.
      If the test organisms were fed while in the laboratory, the data
base for the study should include documentation that the food was
free of residues of chemicals that could adversely affect the
results of the test. There should be a complete record of these
analyses in the archives.

Water Quality

Another area of concern to an aquatic biologist is the quality of
the water being used to hold the organisms during rearing, acclima-
tion and testing. Frequently biologists or technicians can conduct
tests to document pH, dissolved oxygen, alkalinity, hardness, and
salinity in the incoming water and test water. However, it frequently requires a real commitment of the chemistry support to obtain
detailed analyses of the toxic chemical and minerals in the incoming
water used to hold and/or test aquatic organisms. Many of the analyses should be conducted on the incoming water at least semi-annu-
ally (1). In the case of new water supplies, these analyses should
be conducted monthly or quarterly until the laboratory staff is able
to document that no seasonal or periodic changes in water quality
occur that could adversely affect test results. This historical
database should be retained by the laboratory as raw data.

The Agency's scientific staff will evaluate the effect the chemistry findings might have on the outcome of the study.

Potable water containing chlorine or copper is generally considered toxic to many invertebrates. Yet, the lack of minerals, etc., in distilled water makes it osmotically unacceptable to many aquatic organisms. The fact that the water is fit for human consumption does not mean the water is acceptable to aquatic organisms.

Chemical Stability

In some cases it is necessary to analyze the water sample immediately at the test facility because of the rapid breakdown of the chemical in water, particularly at some of the low concentrations that might be toxic to the test organism. If samples are to be stored for any length of time, or shipped to another facility for analysis, the laboratory and/or sponsor should be able to document the test material stability in water at the concentrations being used in the bioassay. The data should be available to the testing facility before a test is initiated. Inadequate documentation of the stability of the chemical under test conditions could result in the study being unsatisfactory for the purpose intended. Agency inspectors should document the availability of the stability data when water and tissue samples are taken for residue analyses during the study.

When analyses on unstable test solutions are not conducted in a timely fashion, the test results are unusable.

Many laboratories have the test chemical examined by the chemistry staff before it is tested in the laboratory. The chemists frequently indicate the solvents to be used, the relative solubility of the test material, and any special instructions concerning the handling or analyses of the test material or test solutions. An early involvement of a chemist in the conduct of the bioassay can save the biologist many hours of wasted effort in trying to prepare test solutions or in analyzing them for residues.

Some very poor handling of very toxic chemicals occurs because the staff is not aware of the toxicity of the test chemical. At some laboratories, excessive safety procedures involving nontoxic chemicals are used because the staff do not know anything about the test material. The laboratory staff should be knowledgeable about any chemical they are testing. For no other reason, they should know the characteristics of the chemical before the treated water is released from the facility.

Literature Cited

1. U.S. Environmental Protection Agency, Committee on Methods for Toxicity Tests with Aquatic Organisms. 1975. Methods for acute toxicity tests with fish, macroinvertebrates, and amphibians. Ecol. Res. Ser. EPA-660/3-75-009.

RECEIVED January 29, 1988

Chapter 19

Proposed Federal Insecticide, Fungicide, and Rodenticide Act

Generic Good Laboratory Practice Standards

Willa Y. Garner[1] and Maureen S. Barge[2]

[1]U.S. Environmental Protection Agency, EN-342,
Washington, DC 20460
[2]FMC Corporation, Princeton, NJ 08543

In evaluating the papers from the symposium for
publication in this book, we, as editors, felt that,
in order to present a comprehensive picture, we
should publish the proposed edition of the FIFRA Good
Laboratory Practice Standards. We incorporated the
proposed changes in the November 29, 1983, Federal
Register publication of the final rule to present
a complete document. The Introduction, Economic
Analysis, Statutory Requirements, and Other Regulatory
Requirements from the proposed rule have not been
included in this document in an effort to conserve
its length and make it more of a working document for
the reader.

ENVIRONMENTAL PROTECTION AGENCY

OFFICE OF PESTICIDES AND TOXIC SUBSTANCES

40 CFR PART 160

[OPP-300165; FRL 3245-5]

FEDERAL INSECTICIDE, FUNGICIDE AND RODENTICIDE ACT (FIFRA);
GOOD LABORATORY PRACTICE STANDARDS

AGENCY: Environmental Protection Agency (EPA).

ACTION: Proposed Rule.

SUMMARY: EPA is proposing to expand the scope of the FIFRA Good
Laboratory Practice (GLP) Standards by requiring GLP compliance for

testing conducted in the field and for such disciplines of testing as ecological effects, chemical fate, residue chemistry, and, as required by 40 CFR 158.160, product performance (efficacy testing). EPA is proposing this amendment in order to ensure the quality and integrity of all data submitted to the Agency in conjunction with pesticide product registration, or other marketing and research permits. EPA is also proposing to amend the FIFRA GLPs to incorporate many of the changes made by the Food and Drug Administration (FDA) to its GLP regulations.

PART 160 - GOOD LABORATORY PRACTICE STANDARDS

Subpart A - General Provisions

Sec.
160.1   Scope.
160.3   Definitions.
160.10   Applicability to studies performed under grants and
         contracts.
160.12   Statement of compliance or non-compliance.
160.15   Inspection of a testing facility.
160.17   Effects of non-compliance.

Subpart B - Organization and Personnel

160.29   Personnel.
160.31   Testing facility management.
160.33   Study director.
160.35   Quality assurance unit.

Subpart C - Facilities

160.41   General.
160.43   Test system care facilities.
160.45   Test system supply facilities.
160.47   Facilities for handling test, control, and reference
         substances.
160.49   Laboratory operation areas.
160.51   Specimen and data storage facilities.

Subpart D - Equipment

160.61   Equipment design.
160.63   Maintenance and calibration of equipment.

Subpart E - Testing Facilities Operation

160.81   Standard operating procedures.
160.83   Reagents and solutions.
160.90   Animal and other test system care.

Subpart F - Test, Control, and Reference Substances

160.105 Test, control, and reference substance characterization.
160.107 Test, control, and reference substance handling.
160.113 Mixtures of substances with carriers.

Subpart G - Protocol for and Conduct of a Study

160.120   Protocol.
160.130   Conduct of a study.
160.135   Physical and chemical characterization studies.

Subparts H-I [Reserved]

Subpart J - Records and Reports

160.185   Reporting of study results.
160.190   Storage and retrieval of records and data.
160.195   Retention of records.

Subpart K - [Reserved]

AUTHORITY: 7 U.S.C. 136 a, 136 c, 136 d, 136 f, 136 j, 136 t,
136 v, 136 w; 21 U.S.C. 346 a, 348, 371; Reorganization Plan No. 3
of 1970.

Subpart A - General Provisions

§ 160.1   Scope

(a) This part prescribes good laboratory practices for
conducting studies that support or are intended to support
applications for research or marketing permits for pesticide
products regulated by the EPA. This part is intended to assure
the quality and integrity of data submitted pursuant to sections
3, 5, 8, 18, and 24(c) of the Federal Insecticide, Fungicide,
and Rodenticide Act (FIFRA) (7 U.S.C. 136a, 136c, 136f, 136q,
and 136v(c)) and sections 408 and 409 of the Federal Food,
Drug, and Cosmetic Act (FFDCA) (21 U.S.C. 346a, 348).

§ 160.3   Definitions.

As used in this part the following terms shall have the
meanings specified:
"Application for research or marketing permit" includes:
(1) An application for registration, amended registration,
or reregistration of a pesticide product under FIFRA sections 3
or 24(c).
(2) An application for an experimental use permit under
FIFRA section 5.
(3) An application for an exemption under FIFRA section 18.
(4) A petition or other request for establishment or modification of a tolerance, for an exemption for the need for a
tolerance, or for other clearance under FFDCA section 408.

(5) A petition or other request for establishment or modification of a food additive regulation or other clearance by EPA under FFDCA section 409.

(6) A submission of data in response to a notice issued by EPA under FIFRA section 3(c)(2)(B).

(7) Any other application, petition, or submission sent to EPA intended to persuade EPA to grant, modify, or leave unmodified a registration or other approval required as a condition of sale or distribution of a pesticide.

"Batch" means a specific quantity or lot of a test or control substance that has been characterized according to § 160.105(a).

"Carrier" means any material (e.g., feed, water, soil, nutrient media) with which the test substance is combined for administration to test organisms.

"Control substance" means any chemical substance or mixture or any other material other than a test substance, feed or water that is administered to the test system in the course of a study for the purpose of establishing a basis for comparison with the test substance for no-effect levels.

"EPA" means the U.S. Environmental Protection Agency.

"Experimental start date" means the first date the test substance is applied to the test system.

"Experimental termination date" means the last date on which data are collected directly from the study.

"FDA" means the U.S. Food and Drug Administration.

"FFDCA" means the Federal Food, Drug, and Cosmetic Act, as amended (21 U.S.C. 321 et seq.).

"FIFRA" means the Federal Insecticide, Fungicide, and Rodenticide Act (7 U.S.C. 136 et seq.).

"Person" includes an individual, partnership, corporation, association, scientific or academic establishment, government agency, or organizational unit thereof, and any other legal entity.

"Quality assurance unit" means any person or organizational element, except the study director, designated by testing facility management to perform the duties relating to quality assurance of the studies.

"Raw data" means any laboratory worksheets, records, memoranda, notes, or exact copies thereof, that are the result of original observations and activities of a study and are necessary for the reconstruction and evalution of the report of that study. In the event that exact transcripts of raw data have been prepared (e.g., tapes which have been transcribed verbatim, dated, and verified accurate by signature), the exact copy or exact transcript may be substituted for the original source as raw data. "Raw data" may include photographs, microfilm or microfiche copies, computer printouts, magnetic media, including dictated observations, and recorded data from automated instruments.

"Reference substance" means any chemical substance or mixture or material other than a test substance, feed, or water that is administered to or used in analyzing the test system in the course of a study for purposes of establishing a basis for comparison with the test substance for known effect levels.

"Specimen" means any material derived from a test system for examination or analysis.

"Sponsor" means:

(1) A person who initiates and supports, by provision of financial or other resources, a study;

(2) A person who submits a study to the EPA in support of an application for a research or marketing permit; or

(3) A testing facility, if it both initiates and actually conducts the study.

"Study" means any experiment in which a test substance is studied in a test system under laboratory conditions or in the environment to determine or help predict its metabolism, product performance (efficacy as required by 40 CFR 158.160), environmental and chemical fate, persistence and residue, or other characteristics in humans, other living organisms, or media. The term does not include basic exploratory studies carried out to determine whether a test substance has any potential utility.

"Study completion date" means the date the final report is signed by the study director.

"Study director" means the individual responsible for the overall conduct of a study.

"Study initiation date" means the date the protocol is signed by the study director.

"Test substance" means a substance or mixture administered or added to a test system in a study, which substance or mixture:

(1) Is the subject of an application for a research or marketing permit supported by the study, or is the contemplated subject of such an application; or

(2) Is an ingredient, impurity, degradation product, metabolite, or radioactive isotope of a substance described by paragraph (1) of this definition, or some other substance related to a substance described by that paragraph, which is used in the study to assist in characterizing the toxicity, metabolism, or other characteristics of a substance described by that paragraph.

"Test system" means any animal, plant, microorganism, chemical or physical matrix (e.g., soil or water), or subparts thereof, to which the test or control substance is administered or added for study. "Test system" also includes appropriate groups or components of the system not treated with the test, control, or reference substance.

"Testing facility" means a person who actually conducts a study, i.e., actually uses the test substance in a test system. "Testing facility" encompasses only those operational units that are being or have been used to conduct studies.

"Vehicle" means any agent which facilitates the mixture, dispersion, or solubilization of a test substance with a carrier.

§ 160.10 Applicability to studies performed under grants and contracts.

When a sponsor or other person utilizes the services of a consulting laboratory, contractor, or grantee to perform all or a part of a study to which this part applies, it shall notify the consulting laboratory, contractor, or grantee that the service is, or is part of, a study that must be conducted in compliance with the provisions of this part.

§ 160.12 Statement of compliance or non-compliance.

Any person who submits to EPA an application for a research or marketing permit and who, in connection with the application, submits data from a study to which this part applies shall include in the application a true and correct statement, signed by the applicant, the sponsor, and the study director, of one of the following types:
(a) A statement that the study was conducted in accordance with this part; or
(b) A statement describing in detail all differences between the practices used in the study and those required by this part; or
(c) A statement that the person was not a sponsor of the study, did not conduct the study, and does not know whether the study was conducted in accordance with this part.

§ 160.15 Inspection of a testing facility.

(a) A testing facility shall permit an authorized employee or duly designated representative of EPA or FDA, at reasonable times and in a reasonable manner, to inspect the facility and to inspect (and in the case of records also to copy) all records and specimens required to be maintained regarding studies to which this part applies. The records inspection and copying requirements should not apply to quality assurance unit records of findings and problems, or to actions recommended and taken, except that EPA may seek production of these records in litigation or formal adjudicatory hearings.
(b) EPA will not consider reliable for purposes of supporting an application for research or marketing permit any data developed by a testing facility or sponsor that refuses to permit inspection in accordance with this part. The determination that a study will not be considered in support of an application for a research or marketing permit does not, however, relieve the applicant for such a permit of any obligation under any applicable statute or regulation to submit the results of the study to EPA.

§ 160.17 Effects of non-compliance.

(a) EPA may refuse to consider reliable for purposes of supporting an application for a research or marketing permit any data from a study which was not conducted in accordance with this part.

(b) Submission of a statement required by § 160.12 which is false may form the basis for cancellation, suspension, or modification of the research or marketing permit, or denial or disapproval of an application for such a permit, under FIFRA sections 3, 5, 6, 18, or 24 or FFDCA sections 408 or 409, or for criminal prosecution under 18 U.S.C. 2 or 1001 or FIFRA section 14, or for imposition of civil penalties under FIFRA section 14.

Subpart B -- Organization and Personnel

§ 160.29   Personnel.

(a) Each individual engaged in the conduct of or responsible for the supervision of a study shall have education, training, and experience, or combination thereof, to enable that individual to perform the assigned functions.

(b) Each testing facility shall maintain a current summary of training and experience and job description for each individual engaged in or supervising the conduct of a study.

(c) There shall be a sufficient number of personnel for the timely and proper conduct of the study according to the protocol.

(d) Personnel shall take necessary personal sanitation and health precautions designed to avoid contamination of test, control, and reference substances and test systems.

(e) Personnel engaged in a study shall wear clothing appropriate for the duties they perform.   Such clothing shall be changed as often as necessary to prevent microbiological, radiological, or chemical contamination of test systems and test, control, and reference substances.

(f) Any individual found at any time to have an illness that may adversely affect the quality and integrity of the study shall be excluded from direct contact with test systems, and test, control, and reference substances, and any other operation or function that may adversely affect the study until the condition is corrected.   All personnel shall be instructed to report to their immediate supervisors any health or medical conditions that may reasonably be considered to have an adverse effect on a study.

§ 160.31   Testing facility management.

For each study, testing facility management shall:

(a) Designate a study director as described in § 160.33 before the study is initiated.

(b) Replace the study director promptly if it becomes necessary to do so during the conduct of a study.

(c) Assure that there is a quality assurance unit as described in § 160.35.

(d) Assure that test and control substances or mixtures have been appropriately tested for identity, strength, purity, stability, and uniformity, as applicable.

(e) Assure that personnel, resources, facilities, equipment, materials and methodologies are available as scheduled.

(f) Assure that personnel clearly understand the functions they are to perform.

(g) Assure that any deviations from these regulations reported by the quality assurance unit are communicated to the study director and corrective actions are taken and documented.

§ 160.33 Study director.

For each study, a scientist or other professional of appropriate education, training, and experience, or combination thereof, shall be identified as the study director. The study director has overall responsibility for the technical conduct of the study, as well as for the interpretation, analysis, documentation, and reporting of results, and represents the single point of study control. The study director shall assure that:

(a) The protocol, including any change, is approved as provided by § 160.120 and is followed.

(b) All experimental data, including observations of unanticipated responses of the test system are accurately recorded and verified.

(c) Unforeseen circumstances that may affect the quality and integrity of the study are noted when they occur, and corrective action is taken and documented.

(d) Test systems are as specified in the protocol.

(e) All applicable good laboratory practice regulations are followed.

(f) All raw data, documentation, protocols, specimens, and final reports are transferred to the archives during or at the close of the study.

§160.35 Quality assurance unit.

(a) A testing facility shall have a quality assurance unit which shall be responsible for monitoring each study to assure management that the facilities, equipment, personnel, methods, practices, records, and controls are in conformance with the regulations in this part. For any given study the quality assurance unit shall be entirely separate from and independent of the personnel engaged in the direction and conduct of that study.

(b) The quality assurance unit shall:

(1) Maintain a copy of a master schedule sheet of all studies conducted at the testing facility indexed by test substance and containing the test system, nature of study, date study was initiated, current status of each study, identity of the sponsor, and the name of the study director.

(2) Maintain copies of all protocols pertaining to all studies for which the unit is responsible.

(3) Inspect each study at intervals adequate to ensure the integrity of the study and maintain written and properly signed records of each periodic inspection showing the date of the inspection, the study inspected, the phase or segment of the study inspected, the person performing the inspection,

findings and problems, action recommended and taken to resolve existing problems, and any scheduled date for reinspection. Any problems which are likely to affect study integrity found during the course of an inspection shall be brought to the attention of the study director and management immediately.

(4) Periodically submit to management and the study director written status reports on each study, noting any problems and the corrective actions taken.

(5) Determine that no deviations from approved protocols or standard operating procedures were made without proper authorization and documentation.

(6) Review the final study report to assure that such report accurately describes the methods and standard operating procedures, and that the reported results accurately reflect the raw data of the study.

(7) Prepare and sign a statement to be included with the final study report which shall specify the dates inspections were made and findings reported to management and to the study director.

(c) The responsibilities and procedures applicable to the quality assurance unit, the records maintained by the quality assurance unit, and the method of indexing such records shall be in writing and shall be maintained. These items including inspection dates, the study inspected, the phase or segment of the study inspected, and the name of the individual performing the inspection shall be made available for inspection to authorized employees or duly designated representatives of EPA or FDA.

(d) An authorized employee or a duly designated representative of EPA or FDA shall have access to the written procedures established for the inspection and may request testing facility management to certify that inspections are being implemented, performed, documented and followed-up in accordance with this paragraph.

Subpart C - Facilities

§ 160.41   General.

Each testing facility shall be of suitable size and construction to facilitate the proper conduct of studies. Testing facilities which are not located within an indoor controlled environment shall be of suitable location to facilitate the proper conduct of studies. Testing facilities shall be designed so that there is a degree of separation that will prevent any function or activity from having an adverse effect on the study.

§160.43   Test system care facilities.

(a) A testing facility shall have a sufficient number of animal rooms or other test system areas, as needed, to ensure: proper separation of species or test systems, isolation of individual projects, quarantine or isolation of animals or other test systems, and routine or specialized housing of animals or other test systems.

(1) In tests with plants or aquatic animals, proper separation of species can be accomplished within a room or area by housing them separately in different chambers or aquaria. Separation of species is unnecessary where the protocol specifies the simultaneous exposure of two or more species in the same chamber, aquarium, or housing unit.

(2) Aquatic toxicity tests for individual projects shall be isolated to the extent necessary to prevent cross-contamination of different chemicals used in different tests.

(b) A testing facility shall have a number of animal rooms or other test system areas separate from those described in paragraph (a) of this section to ensure isolation of studies being done with test systems or test, control, and reference substances known to be biohazardous, including volatile substances, aerosols, radioactive materials, and infectious agents.

(c) Separate areas shall be provided, as appropriate, for the diagnosis, treatment, and control of laboratory test system diseases. These areas shall provide effective isolation for the housing of test systems either known or suspected of being diseased, or of being carriers of disease, from other test systems.

(d) Facilities shall have proper provisions for collection and disposal of contaminated water, soil, or other spent materials. When animals are housed, facilities shall exist for the collection and disposal of all animal waste and refuse or for safe sanitary storage of waste before removal from the testing facility. Disposal facilities shall be so provided and operated as to minimize vermin infestation, odors, disease hazards, and environmental contamination.

(e) Facilities shall have provisions to regulate environmental conditions (e.g., temperature, humidity, photoperiod) as specified in the protocol.

(f) For marine test organisms, an adequate supply of clean sea water or artificial sea water (prepared from deionized or distilled water and sea salt mixture) shall be available. The ranges of composition shall be as specified in the protocol.

(g) For freshwater organisms, an adequate supply of clean water of the appropriate hardness, pH, and temperature, and free of contaminants capable of interfering with the study, shall be available as specified in the protocol.

(h) For plants, an adequate supply of soil of the appropriate composition, as specified in the protocol, shall be available as needed.

§ 160.45   Test system supply facilities.

(a) There shall be storage areas, as needed, for feed, nutrients, soils, bedding, supplies, and equipment. Storage areas for feed, nutrients, soils, and bedding shall be separated from areas housing the test systems and shall be protected against infestation or contamination. Perishable supplies shall be preserved by appropriate means.

(b) When appropriate, plant supply facilities shall be provided. These include:
(1) Facilities, as specified in the protocol, tor holding, culturing, and maintaining algae and aquatic plants.
(2) Facilities, as specified in the protocol, for plant growth (e.g., greenhouses, growth chambers, light banks).
(c) When appropriate, facilities for aquatic animal tests shall be provided. These include aquaria, holding tanks, ponds, and ancillary equipment, as specified in the protocol.

§ 160.47 Facilities for handling test, control, and reference substances.

(a) As necessary to prevent contamination or mixups, there shall be separate areas for:
(1) Receipt and storage of the test, control, and reference substances.
(2) Mixing of the test, control, and reference substances with a carrier, e.g., feed.
(3) Storage of the test, control, and reference substance mixtures.
(b) Storage areas for the test, control, and/or reference substance and for test, control, and/or reference mixtures shall be separate from areas housing the test systems and shall be adequate to preserve the identity, strength, purity, and stability of the substances and mixtures.

§ 160.49 Laboratory operation areas.

Separate laboratory space and other space shall be provided, as needed, for the performance of the routine and specialized procedures required by studies.

§ 160.51 Specimen and data storage facilities.

Space shall be provided for archives, limited to access by authorized personnel only, for the storage and retrieval of all raw data and specimens from completed studies.

Subpart D - Equipment

§ 160.61 Equipment design.

Equipment used in the generation, measurement, or assessment of data and equipment used for facility environmental control shall be of appropriate design and adequate capacity to function according to protocol and shall be suitably located for operation, inspection, cleaning, and maintenance.

§ 160.63 Maintenance and calibration of equipment.

(a) Equipment shall be adequately inspected, cleaned, and maintained. Equipment used for the generation, measurement, or assessment of data shall be adequately tested, calibrated, and/or standardized.

Published literature may be used as a supplement to standard operating procedures.

(d) A historical file of standard operating procedures, and all revisions thereof, including the dates of such revisions, shall be maintained.

§ 160.83   Reagents and solutions.

All reagents and solutions in the laboratory areas shall be labeled to indicate identity, titer or concentration, storage requirements, and expiration date. Deteriorated or outdated reagents and solutions shall not be used.

§ 160.90   Animal and other test system care.

(a) There shall be standard operating procedures for the housing, feeding, handling, and care of animals and other test systems.

(b) All newly received test systems from outside sources shall be isolated and their health status or appropriateness for the study evaluated. This evaluation shall be in accordance with acceptable veterinary medical practice or scientific practice.

(c) At the initiation of a study, test systems shall be free of any disease or condition that might interfere with the purpose or conduct of the study. If, during the course of the study, the test systems contract such a disease or condition, the diseased test systems should be isolated, if necessary. These test systems may be treated for disease or signs of disease provided that such treatment does not intefere with the study. The diagnosis, authorization of treatment, description of treatment, and each date of treatment shall be documented and shall be retained.

(d) Warm-blooded animals, adult reptiles, and adult terrestrial amphibians used in laboratory procedures that require manipulations and observations over an extended period of time or in studies that require these test systems to be removed from and returned to their test system-housing units for any reason (e.g., cage cleaning, treatment, etc.) shall receive appropriate identification (e.g., tattoo, toe clip, color code, ear tag, ear punch, etc.). All information needed to specifically identify each test system within the test system-housing unit shall appear on the outside of that unit. Suckling mammals and juvenile birds are excluded from the requirement of individual identification unless otherwise specified in the protocol.

(e) Except as specified in paragraph (e)(1) of this section, test systems of different species shall be housed in separate rooms when necessary. Test systems of the same species, but used in different studies, should not ordinarily be housed in the same room when inadvertent exposure to test, control, or reference substances or test system mixup could affect the outcome of either study. If such mixed housing is necessary, adequate differentiation by space and identification shall be made.

(1) Plants, invertebrate animals, aquatic vertebrate animals, and organisms that may be used in multispecies tests need

not be housed in separate rooms, provided that they are adequately segregated to avoid mixup and cross contamination.

(2) [Reserved]

(f) Cages, racks, pens, enclosures, aquaria, holding tanks, ponds, growth chambers, and other holding, rearing and breeding areas, and accessory equipment, shall be cleaned and sanitized at appropriate intervals.

(g) Feed, soil, and water used for the test systems shall be analyzed periodically to ensure that contaminants known to be capable of interfering with the study and reasonably expected to be present in such feed, soil, or water are not present at levels above those specified in the protocol. Documentation of such analyses shall be maintained as raw data.

(h) Bedding used in animal cages or pens shall not interfere with the purpose or conduct of the study and shall be changed as often as necessary to keep the animals dry and clean.

(i) If any pest control materials are used, the use shall be documented. Cleaning and pest control materials that interfere with the study shall not be used.

(j) All plant and animal test organisms shall be acclimatized, prior to their use in an experiment, to the environmental conditions of the test.

Subpart F - Test, Control, and Reference Substances

§ 160.105   Test, control, and reference substance characterization.

(a) The identity, strength, purity, and composition or other characteristics which will appropriately define the test, control, or reference substance shall be determined for each batch and shall be documented before its use in an experiment. Methods of synthesis, fabrication, or derivation of the test, control, or reference substance shall be documented by the sponsor or the testing facility.

(b) The stability and, when relevant to the conduct of the experiment, the solubility of each test, control, or reference substance shall be determined by the testing facility or by the sponsor before the experimental start date. Where periodic analysis of each batch is required by the protocol, there shall be written standard operating procedures that shall be followed.

(c) Each storage container for a test, control, or reference substance shall be labeled by name, chemical abstracts service (CAS) number or code number, batch number, expiration date, if any, and, where appropriate, storage conditions necessary to maintain the identity, strength, purity, and composition of the test, control, or reference substance. Storage containers shall be assigned to a particular test substance for the duration of the study.

(d) For studies of more than 4 weeks' duration, reserve samples from each batch of test, control, and reference substance shall be retained for the period of time provided by § 160.195.

(e) The stability of test, control, and reference substances under test conditions shall be known for all studies.

§ 160.107   Test, control, and reference substance handling.

Procedures shall be established for a system for the handling of the test, control, and reference substances to ensure that:
(a) There is proper storage.
(b) Distribution is made in a manner designed to preclude the possibility of contamination, deterioration, or damage.
(c) Proper identification is maintained throughout the distribution process.
(d) The receipt and distribution of each batch is documented. Such documentation shall include the date and quantity of each batch distributed or returned.

§ 160.113   Mixtures of substances with carriers.

(a) For each test, control, or reference substance that is mixed with a carrier, tests by appropriate analytical methods shall be conducted:
(1) To determine the uniformity of the mixture and to determine, periodically, the concentration of the test, control, or reference substance in the mixture.
(2) To determine the stability and, when relevant to the conduct of the experiment, the solubility of the test, control, or reference substance in the mixture before the experimental start date. Determination of the stability and solubility of the test, control, or reference substance in the mixture shall be done under the environmental conditions specified in the protocol and as required by the conditions of the experiment. Where periodic analysis of the mixture is required by the protocol, there shall be written standard operating procedures that shall be followed.
(b) Where any of the components of the test, control, or reference substance carrier mixture has an expiration date, that date shall be clearly shown on the container. If more than one component has an expiration date, the earliest date shall be shown.
(c) If a vehicle is used to facilitate the mixing of a test substance with a carrier, assurance shall be provided that the vehicle does not interfere with the integrity of the test.

Subpart G – Protocol for and Conduct of a Study

§ 160.120 Protocol.

(a) Each study shall have an approved written protocol that clearly indicates the objectives and all methods for the conduct of the study. The protocol shall contain but shall not necessarily be limited to the following information:
(1) A descriptive title and statement of the purpose of the study.
(2) Identification of the test, control, and reference substance by name, chemical abstracts service (CAS) number or code number.
(3) The name and address of the sponsor and the name and address of the testing facility at which the study is being conducted.

(4) The proposed experimental start and termination dates.

(5) Justification for selection of the test system.

(6) Where applicable, the number, body weight, sex, source of supply, species, strain, substrain, and age of the test system.

(7) The procedure for identification of the test system.

(8) A description of the experimental design, including methods for the control of bias.

(9) Where applicable, a description and/or identification of the diet used in the study as well as solvents, emulsifiers and/or other materials used to solubilize or suspend the test, control, or reference substances before mixing with the carrier. The description shall include specifications for acceptable levels of contaminants that are reasonably expected to be present in the dietary materials and are known to be capable of interfering with the purpose or conduct of the study if present at levels greater than established by the specifications.

(10) The route of administration and the reason for its choice.

(11) Each dosage level, expressed in milligrams per kilogram of body or test system weight or other appropriate units, of the test, control, or reference substance to be administered and the method and frequency of administration.

(12) The type and frequency of test analyses, and measurements to be made.

(13) The records to be maintained.

(14) The date of approval of the protocol by the sponsor and the dated signature of the study director.

(15) A statement of the proposed statistical method.

(b) All changes in or revisions of an approved protocol and the reasons therefore shall be documented, signed by the study director, dated, and maintained with the protocol.

§ 160.130 Conduct of a study.

(a) The study shall be conducted in accordance with the protocol.

(b) The test systems shall be monitored in conformity with the protocol.

(c) Specimens shall be identified by test system, study, nature, and date of collection. This information shall be located on the specimen container or shall accompany the specimen in a manner that precludes error in the recording and storage of data.

(d) In animal studies where histopathology is required, records of gross findings for a specimen from postmortem observations shall be available to a pathologist when examining that specimen histopathologically.

(e) All data generated during the conduct of a study, except those that are generated by automated data collection systems, shall be recorded directly, promptly, and legibly in ink. All data entries shall be dated on the day of entry and signed or initialed by the person entering the data. Any change in entries shall be made so as not to obscure the original entry, shall indicate the reason for such change, and shall be dated and signed or identified at the time of the change. In automated data collection systems, the individual responsible for direct data input shall be identified at the time of data input. Any change in automated data entries shall be made so as not to obscure the original entry, shall indicate the reason for change, shall be dated, and the responsible individual shall be identified.

§ 160.135   Physical and chemical characterization studies.

(a) Except as provided in paragraph (b) of this section, the following provisions shall not apply to studies designed to determine physical and chemical characteristics of a test, control, or reference substance:
§ 160.31(c), (d), and (g)
§ 160.35(b) and (c)
§ 160.43
§ 160.45
§ 160.47
§ 160.49
§ 160.81(b)(1), (2), (6) through (9), and (12)
§ 160.90
§ 160.105(a) through (d)
§ 160.113
§ 160.120(a)(5) through (12), and (15)
§ 160.185(a)(5) through (8), (10), (12), and (14)
§ 160.195(c) and (d).

(b) The exemptions provided in paragraph (a) of this section shall not apply to physical/chemical characterization studies designed to determine stability, solubility, octanol water partition coefficient, volatility, and persistence (such as bio-degradation, photodegradation, and chemical degradation studies), and such studies shall be conducted in accordance with this part.

Subparts H and I - [Reserved]

Subpart J - Records and Reports

§ 160.185   Reporting of study results.

(a) A final report shall be prepared for each study and shall include, but not necessarily be limited to, the following:
(1) Name and address of the facility performing the study and the dates on which the study was initiated and was completed, terminated, or discontinued.
(2) Objectives and procedures stated in the approved protocol, including any changes in the original protocol.
(3) Statistical methods employed for analyzing the data.
(4) The test, control, and reference substances identi-fied by name, chemical abstracts service (CAS) number or code number, strength, purity, and composition, or other appropriate characteristics.
(5) Stability and, when relevant to the conduct of the experiment, the solubility of the test, control, and reference substances under the conditions of administration.
(6) A description of the methods used.

(7) A description of the test system used. Where applicable, the final report shall include the number of animals used, sex, body weight range, source of supply, species, strain and substrain, age, and procedure used for identification.

(8) A description of the dosage, dosage regimen, route of administration, and duration.

(9) A description of all circumstances that may have affected the quality or integrity of the data.

(10) The name of the study director, the names of other scientists or professionals, and the names of all supervisory personnel involved in the study.

(11) A description of the transformations, calculations, or operations performed on the data, a summary and analysis of the data, and a statement of the conclusions drawn from the analysis.

(12) The signed and dated reports of each of the individual scientists or other professionals involved in the study, including each person who, at the request or direction of the testing facility or sponsor, conducted an analysis or evaluation of data or specimens from the study after data generation was completed.

(13) The locations where all specimens, raw data, and the final report are to be stored.

(14) The statement prepared and signed by the quality assurance unit as described in § 160.35(b)(7).

(b) The final report shall be signed and dated by the study director.

(c) Corrections or additions to a final report shall be in the form of an amendment by the study director. The amendment shall clearly identify that part of the final report that is being added to or corrected and the reasons for the correction or addition, and shall be signed and dated by the person responsible.

(d) A copy of the final report and of any amendment to it shall be maintained by the sponsor and the testing facility.

§ 160.190 Storage and retrieval of records and data.

(a) All raw data, documentation, records, protocols, specimens, and final reports generated as a result of a study shall be retained. Specimens obtained from mutagenicity tests, specimens of soil, water, and plants, and wet specimens of blood, urine, feces, and biological fluids do not need to be retained beyond quality assurance. Correspondence and other documents relating to interpretation and evaluation of data, other than those documents contained in the final report, also shall be retained.

(b) There shall be archives for orderly storage and expedient retrieval of all raw data, documentation, protocols, specimens, and interim and final reports. Conditions of storage shall minimize deterioration of the documents or specimens in accordance with the requirements for the time period of their retention and the nature of the documents or specimens. A testing facility may contract with commercial archives to provide a repository for all materials to be retained. Raw data and specimens may be retained elsewhere provided that the archives have specific reference to those other locations.

(c) An individual shall be identified as responsible for the archives.

(d) Only authorized personnel shall enter the archives.

(e) Material retained or referred to in the archives shall be indexed to permit expedient retrieval.

§ 160.195 Retention of records.

(a) Record retention requirements set forth in this section do not supersede the record retention requirements of any other regulations in this subchapter.

(b) Except as provided in paragraph (c) of this section, documentation records, raw data, and specimens pertaining to a study and required to be retained by this part shall be retained in the archive(s) for whichever of the following periods is longest:

(1) In the case of any study used to support an application for a research or marketing permit approved by EPA, the period during which the sponsor holds any research or marketing permit to which the study is pertinent.

(2) A period of at least five years following the date on which the results of the study are submitted to the EPA in support of an application for a research or marketing permit.

(3) In other situations (e.g., where the study does not result in the submission of the study in support of an application for a research or marketing permit), a period of at least two years following the date on which the study is completed, terminated, or discontinued.

(c) Wet specimens, samples of test, control, or reference substances, and specially prepared material which are relatively fragile and differ markedly in stability and quality during storage, shall be retained only as long as the quality of the preparation affords evaluation. Specimens obtained from mutagencity tests, specimens of soil, water, and plants, and wet specimens of blood, urine, feces, biological fluids, do not need to be retained beyond quality assurance review. In no case shall retention be required for longer periods than those set forth in paragraph (b) of this section.

(d) The master schedule sheet, copies of protocols, and records of quality assurance inspections, as required by § 160.35(c) shall be maintained by the quality assurance unit as an easily accessible system of records for the period of time specified in paragraph (b) of this section.

(e) Summaries of training and experience and job descriptions required to be maintained by § 160.29(b) may be retained along with all other testing facility employment records for the length of time specified in paragraph (b) of this section.

(f) Records and reports of the maintenance and calibration and inspection of equipment, as required by § 160.63(b) and (c), shall be retained for the length of time specified in paragraph (b) of this section.

(g) If a facility conducting testing or an archive contracting facility goes out of business, all raw data, documentation, and other material specified in this section shall be transferred to the archives of the sponsor of the study. The EPA shall be notified in writing of such a transfer.

(h) Specimens, samples, or other non-documentary materials need not be retained after EPA has notified in writing the sponsor or testing facility holding the materials that retention is no longer required by EPA. Such notification normally will be furnished upon request after EPA or FDA has completed an audit of the particular study to which the materials relate and EPA has concluded that the study was conducted in accordance with this part.

(i) Records required by this part may be retained either as original records or as true copies such as photocopies, microfilm, microfiche, or other accurate reproduction of the original records.

Subpart K - [Reserved]

Literature Cited

1. Pesticide Programs, Good Laboratory Practice Standards; Final Rule, Fed. Reg., 48:53946, November 29, 1983.

2. Federal Insecticide, Fungicide and Rodenticide Act (FIFRA) and Toxic Substances Control Act (TSCA); Good Laboratory Practice Standards; Proposed Rules, Fed. Reg., 52:48920, December 28, 1987.

RECEIVED January 12, 1988

Author Index

Affiliation Index

Subject Index

Production by Meg Marshall
Indexing by Deborah H. Steiner
Jacket design by Carla L. Clemens

Elements typeset by Hot Type Ltd., Washington, DC
Printed and bound by Maple Press, York, PA

Recent ACS Books

Chemical Demonstrations: A Sourcebook for Teachers
By Lee R. Summerlin and James L. Ealy, Jr.
192 pp; spiral bound; ISBN 0–8412–0923–5

Silent Spring Revisited
Edited by Gino J. Marco, Robert M. Hollingworth, and William Durham
214 pp; clothbound; ISBN 0–8412–0980–4

The ACS Style Guide: A Manual for Authors and Editors
Edited by Janet S. Dodd
264 pp; clothbound; ISBN 0–8412–0917–0

Personal Computers for Scientists: A Byte at a Time
By Glenn I. Ouchi
276 pp; clothbound; ISBN 0–8412–1000–4

Writing the Laboratory Notebook
By Howard M. Kanare
146 pp; clothbound; ISBN 0–8412–0906–5

Principles of Environmental Sampling
Edited by Lawrence H. Keith
458 pp; clothbound; ISBN 0–8412–1173–6

Supercritical Fluid Extraction and Chromatography:
Techniques and Applications
Edited by Bonnie A. Charpentier and Michael R. Sevenants
ACS Symposium Series 366; 253 pp; clothbound; ISBN 0–8412–1469–7

Food and Packaging Interactions
Edited by Joseph H. Hotchkiss
ACS Symposium Series 365; 305 pp; clothbount; ISBN 0–8412–1465–4

Chemical Reactions on Polymers
Edited by Judith L. Benham and James F. Kinstle
ACS Symposium Series 364; 483 pp; ISBN 0–8412–1448–4

Pharmacokinetics: Processes and Mathematics
By Peter G. Welling
ACS Monograph 185; 290 pp; ISBN 0–8412–0967–7

Polynuclear Aromatic Compounds
Edited by Lawrence B. Ebert
Advances in Chemistry Series 217; 396 pp; ISBN 0–8412–1014–4

For further information and a free catalog of ACS books, contact:
American Chemical Society
Distribution Office, Department 225
1155 16th Street, NW, Washington, DC 20036
Telephone 800–227–5558